TRENDS IN GENETIC ANALYSIS

Number XVIII of the Columbia Biological Series

TRENDS IN
GENETIC ANALYSIS

BY G. PONTECORVO, F.R.S.

Professor of Genetics, University of Glasgow

 1958 NEW YORK

COLUMBIA UNIVERSITY PRESS

25341

COLUMBIA BIOLOGICAL SERIES
Edited at Columbia University

CONTENTS

CONTENTS

TABLES

PLATE

TRENDS IN GENETIC ANALYSIS

INTRODUCTION

THIS book is based on the Jesup Lectures delivered in April, 1956, in the Department of Zoology of Columbia University in the City of New York. A portion of the text is unchanged and the colloquial style generally retained.

The title—for short, "Trends in Genetic Analysis"—is an obvious overstatement: there was no intention, of course, of dealing in six lectures with all the directions along which genetics is advancing. The treatment was confined only to those fields with which the author has firsthand acquaintance. Cytoplasmic inheritance, biometrical genetics, and even the detailed study of mutation were completely omitted. Yet no one, least of all the author, will maintain that there are no trends there. These are all frontier fields expanding very vigorously. The scope of the lectures was a reappraisal, on the basis of present knowledge, of the theory of the gene.

A by-product of recent research is the realization that sexual reproduction—i.e., a regular alternation of karyogamy and meiosis as shown in higher organisms—is by no means the only process for the pooling and reassorting of genetic information from different lines of descent. Though known so far only in microorganisms, novel processes of genetic recombination make it clear that some modernized version of the theory of the gene is applicable in organisms or situations in which sexual reproduction (the basis of the original theory) does not occur. The two closing chapters of this book deal precisely with these novel processes.

To put the content of this book in its correct historical perspective, we have to remember a few landmarks. Bateson and Punnett discovered linkage in 1906. In 1912 Morgan and his group at Columbia University correlated the recombination of linked genes with the occurrence of reciprocal exchanges between homologous chromosomes at meiosis, which had been demonstrated by Janssens in 1909. The linear arrangement of genes and the first linkage map by Sturtevant soon followed in 1913. The next landmark is the series of papers by Muller on the mechanism of crossing over, published in 1916. With this work and with that by Bridges on "non-disjunction," all the elements for the theory of the gene had been gathered, though this wording was used by Morgan only in 1917. The theory was expounded by the Columbia University team in 1915 as "The Mechanism of Mendelian heredity." The term "gene," coined by Johannsen in 1911, has already become established.

It is no exaggeration to say that before about 1940 what was known on the nature and the mode of transmission of genetic specificity—*i.e.*, what was known about chromosomal heredity—was but a series of developments of the theory of the gene.

First came the studies of mutation started by Muller about 1920 that led in 1927 to the discovery by Muller and by Stadler of the mutagenic effects of radiations. It took about fifteen years before chemical mutagenesis was demonstrated by Auerbach and Robson and a vast new field opened up.

Second, in the 1920s Wilson, Belling, and later Darlington unified Mendelian inheritance and chromosome cytology. About the same period witnessed another major success: the unification of Mendelism and Darwinism, which the theoretical work of Fisher, Sewall Wright, Haldane, and Tchetverikov made possible. It also witnessed a substantial beginning: the analysis of the effects of genes in terms of biochemical and developmental processes, in which Haldane, Goldschmidt, and later Ephrussi played leading roles.

Third, in the years immediately preceding World War II, something quite new happened: the introduction of ideas (not techniques) from the realm of physics into the realm of genetics, particularly applied to the problems of the size, mutability, and self-replication of genes. The names of Jordan, Frank-Kamenetski, Friedrich-Freska, Zimmer, and Delbrück, with Muller and Timofeef-Ressovsky as their biological interpreters, are linked to this development. Though this first application of physical ideas to a particular set of problems did not work out too well, the whole outlook in theoretical genetics has since been perfused with a physical flavour.

The debt of genetics to physics, and to physical chemistry, for ideas began to be substantial then, and it has been growing steadily all the time. Techniques from physics and physical chemistry, on the other hand, have contributed very little to genetics. This is in sharp contrast to the relations of genetics with chemistry and biochemistry, which have contributed innumerable techniques and facts, but few, if any, ideas.

The historical landscape sketched here is what appeared to the writer in 1939 when, at an age older than is usual for a neophyte, he started learning the elements of genetics. No doubt the picture is blurred and incomplete: some landmarks are forgotten and others are given too much prominence. But the main point, which can be hardly controversial, is the one made above, that up to 1940, or thereabouts, genetics was essentially a development of the "Theory of the Gene." Its impact has been profound, and has remoulded the present ways of looking at living things and investigating the living world. Its impact on politics of the extreme "left" or the extreme "right" has also had far-reaching results, but few would say for the better.

Since about 1940 there has been a gradual change in the outlook in genetics. One reason for this change is the realisation that the theory of the gene, though still indispensable in everyday genetics, is no longer of heuristic value at levels of further refinement, especially when it comes to the enquiry of what the genetic materials are and how they work. The analogy here—and it is almost a platitude—is with classical physics versus quantum mechanics. Unfortunately the quantum mechanics of genetics is not yet with us.

Another reason is technical: the developments brought about by the use of microorganisms in genetics. This has enormously increased the resolving power of genetic analysis and has stimulated very fruitful general ideas from the study of relatively simpler genetic systems.

A third reason is the closer relationship between biochemistry and genetics. For this we are indebted mostly to Beadle and Tatum, who, quite independently, rediscovered Garrod's idea of "inborn errors of metabolism" and applied it to biosyntheses in microorganisms. By doing so they provided a technique of immense and versatile power: suffice it to say that most of the development of microbial genetics is based on this technique. Unfortunately, by and large, this technique has not been put to the best possible use in one of the directions for which it has exceptional value, *i.e.*, the study of the primary actions of the genetic material and their relations to its fine structure. In this respect, it has been made mainly into a tool for the unexciting description of intermediary metabolism, for which it competes or cooperates with half a dozen other more traditional techniques. Only occasionally has it been used for the study of the biochemical system of which the genetic material is one component. It is still full of promise but has not yet made fundamental achievements. The large amount of information collected since 1940 on the genetic control of biosyntheses has so far only descriptive value.

A fourth reason is the impact of ideas from information theory, especially in relation to molecular structure. It is too soon to assess its results, but I have no doubt about its decisive importance.

Clearly, if we are to free ourselves of the fetters of purely formal genetics, of genetics based on abstractions—though valid abstractions—of genetics as merely the mechanics of hereditary transmission, there is no doubt that we have to give physical, chemical, and physiological content to the processes of heredity, variation, and differentiation. We have to express such concepts as gene, allele, mutation, crossing over, dominance, etc., in terms of precise processes taking place in or on structures of the cell.

So far there is a certain reluctance both in approaches and in techniques to giving structure to the chemical processes within the cell: *i.e.*, trying to do something about the fact that spatial arrangement at the megamolecular or even higher levels is an essential part of the game. Coupled reactions on surfaces, relations of molecular structure to biological activity, arrangement of reactants in microvessels, are the things that matter for our purpose, rather than the elegant unravelling of pathways of intermediary metabolism. This is why physicists and physical chemists have contributed decisively to biological thought. This is why attempts like those of Watson and Crick, Pauling, Szent-Gyorgy, and Astbury, have such great appeal.

The fundamental problems in genetics with which we are faced are still the same as those which Muller so clearly stated in his Pilgrim Trust Lecture, "The Gene," at The Royal Society in 1945: among these, the nature of the self-duplicating process of genetic structures, the nature of gene effects, and the nature of the recombination process.

As to the first two, but not the third (see Chapter IV), we can now state the problems more specifically, and there are some testable models. For instance, on the one hand we can now test whether or not the atoms of an existing genetic structure are distributed at random between the two structures resulting from its self-duplication. The brilliant experiments with isotopes by Levinthal (1956) with bacteriophage, and by Plaut and Mazia (1956) and Taylor (1957) with chromosomes, show that they are not.

On the other hand Watson and Crick (1953) and Kacser (1956) have introduced an idea which does not seem to have occurred before (for example, see Muller, 1947a, p. 21), *i.e.*, that the self-duplicating structure itself has a complementary architecture like a positive and a negative photograph face to face.

According to this idea, the duplication process of a structure symbolised as $\overset{*\,*}{AB}$ (* indicates parental structure) consists in $\overset{*}{A}$ combining with B's building blocks to give a "daughter" $\overset{*}{\text{A}}$B, and $\overset{*}{\text{B}}$ combining with A's building blocks to give a "daughter" A$\overset{*}{\text{B}}$.

Watson and Crick applied this idea to the molecular (double helix) structure they proposed for DNA: in this case A and B stand for nucleotides able to pair by hydrogen bonds. Kacser applied it to a protein-DNA complementary architecture of supramolecular size in which the genetic specificity is a property of the interface between the two constituents.

Penrose and Penrose (1957 and unpublished), however, in a note which deserves more consideration than it has received, have shown that it is possible to devise even large mechanical models which are self-duplicating. An essential feature of these models is a symmetrical structure say $\overset{*}{\text{A}}\overset{*}{\text{A}}$, which can duplicate either by accretion on each side followed by splitting $(\text{A}\overset{*}{\text{A}}\overset{*}{\text{A}}\text{A} \rightarrow \text{A}\overset{*}{\text{A}} + \overset{*}{\text{A}}\text{A})$ or on one side only $(\overset{*}{\text{A}}\overset{*}{\text{A}}\text{AA} \rightarrow \overset{*}{\text{A}}\overset{*}{\text{A}} + \text{AA})$.

Models of these kinds have predictable consequences, and it is not too optimistic to believe that experiments of the types already mentioned will soon be able to discriminate between them.

As to the nature of gene effects, Beadle and Tatum's technique has provided material at will. The most promising line seems to be the analysis of cases in which a heterozygote produces both substances, say proteins, each of which is produced by each of the two homozygotes. Horowitz's work on tyrosinases in Neurospora is one of the best examples of this kind.

The identification of the precise difference between the two substances may give a clue as to what the primary gene action is. For instance, Ingram (1957) has shown that the haemoglobin of individuals homozygous for sickle cell anaemia (a gene-determined abnormality) differs from normal in one out of its 300-odd aminoacid residues. The heterozygotes have both types of haemoglobin.

The next fascinating step is that of finding out what it is, in the gene "code," that determines this difference. Again this aim is not as fantastic as it would have been in 1945. It requires the combination of work like that of Ingram on a gene-determined protein with work like that of Benzer (1957) or Pritchard (1955) on the fine structure of a gene.

Certain aspects of these problems of genetics are the main subject of this book.

CHAPTER I

GENETIC ANALYSIS AND ITS RESOLVING POWER

"ANALYSIS," in the Oxford dictionary, is defined as "resolution into simple elements." In genetic analysis we must be clear about what we resolve and into what simpler elements.

Classical genetic analysis is based on the results of breeding and by means of them resolves the genome into linkage groups, and each linkage group into loci. By also making use of cytological techniques and combining them with breeding techniques it goes further: it establishes on which chromosome each linkage group has its structural basis and to which small section of the chromosome each locus corresponds.

Mainly as a consequence of the development of microbial genetics, genetic analysis has increased enormously its resolving power in recent years, so much so that it now goes beyond that of physical or chemical techniques applied to biological organisation. I hope to substantiate this contention and make it more precise than was possible in 1952 when it was first put forward.

The essential process on which genetic analysis is based is recombination. Consider the analogy with microscopy, which is based instead on diffraction. The resolving power attained in microscopy depends on the quality of the microscope and on other technical details, but we know that it has a theoretical limit set by the wavelength of the light used. So far, in genetic analysis the resolving power has been limited only by the refinement of techniques. What the ultimate limit is we do not know, nor can we deduce from theory. Recent advances make it possible to venture a few guesses.

Recombination can be defined as any process which gives origin to cells or individuals associating in new ways two or more hereditary determinants in which their ancestors differed: for instance, cells with determinants Ab or aB descending from other cells with AB or ab.

Until less than fifteen years ago, only two processes of recombination were known: sexual reproduction and infection. Now we know that there are more. For instance, transformation by means of desoxyribonucleic acid (Avery, MacLeod, and McCarthy, 1944) and virus-mediated transduction (Zinder and Lederberg, 1952) in bacteria, the parasexual cycle (Pontecorvo, 1954) in fungi, etc.

We recognise recombination by observing in a line of descent certain cells or individuals—recombinants—which show new associations of properties. Recombination of properties, however, is only the detectable secondary effect of reassociation of subcellular structures determining differences in such properties.

In the type of recombination on which classical genetic analysis is based, $i.e.$, recombination in sexual reproduction, these structures are the chromosomes and their linearly arranged elements. The latter are recognised as genes as a consequence of their specific activities in metabolism and development.

In sexual reproduction recombination of chromosomes and their elements takes place at meiosis and it is the result of the independent segregation of nonhomologous chromosome pairs and of crossing over between members of a chromosome pair, respectively.

Crossing over (whatever its precise mechanism, see Chapter IV) can be formally described as the reciprocal exchange of linear bonds at corresponding positions along pairs of homologous chromosomes. These exchanges are microscopically observable in suitable material. In a population of cells going through meiosis, the incidence of exchanges between any two given points in one chromosome pair is highly correlated with the physical distance between these two points. This incidence can vary from an average of over five exchanges per cell per chromosome pair to none.

In the analysis of the results of breeding experiments we recognise recombination not by microscopic examination of cells in meiosis but by the proportion of recombinant individuals in later generations of a cross. From the fraction of recombinant individuals one can calculate the fraction of recombinant gametes from which those individuals originated. The two coincide, of course, in the case of haploid organisms.

If the proportion of the recombinant gametes formed—e.g., Ab and aB—is smaller than that of gametes with the ancestral associations—AB and ab ("parental")—we say that there is linkage between the two genes A/a and B/b. This is almost the same as saying that along a chromosome pair between the position of the difference A versus a and that of the difference B versus b there is an incidence of less than one exchange as the average of a large number of cells in meiosis.

The measure of linkage is based on the fraction of recombinant gametes out of the total tested. If the two positions just mentioned are either on different pairs of chromosomes or so far apart on one pair that there are on the average one or more exchanges between them, the recombinant gametes are formed in equal proportions with the "parental" gametes. In this case we say that there is "free recombination" between the two genes A/a and B/b.

Three or more genes closely linked two by two (*i.e.*, showing much less than 50% recombination) reveal the additiveness of recombination fractions. If the recombination fraction between A/a and B/b is, say, .05 and that between B/b and C/c is .04, that between A/a and C/c will be approximately either the sum of (.09) or the difference between (.01) the other two. This additiveness makes it possible to represent the recombination fractions graphically as segments in a linear sequence—A/a, B/b, C/c in the first case, or A/a, C/c, B/b in the second case—with the length of the intervals proportional in suitable scale to the recombination fractions.

Overlooking certain refinements required by the occurrence of multiple exchanges, interference, etc., which lead us to correct the recombination fractions and turn them into "crossover values," a linkage map consists essentially of this. The "position" of a gene on the linkage map is called its locus.

Thus, by means of breeding experiments, we measure the incidence of recombination between a group of linked loci and we resolve a linkage group into crossover values between loci. This gives a valid but rather abstract picture. We do not make it much more concrete by representing the crossover values graphically as a linkage map. In the same way we do not make more concrete the tabulation of volumes versus pressures of a gas if we turn it into a graph, though for certain people like me it is easier to grasp the meaning of a graph than that of a table or an equation.

Because of the approximate additiveness of recombination fractions, linkage maps are quite meaningful but are still only graphic expressions of certain numerical relationships. They are more meaningful than the analogy of pressures and volumes because the material structure—the chromosome—which underlies the relations expressed in the map actually has a linear arrangement at least at first approximation.

The concepts and the methods which have led to the construction of linkage maps in organisms with a standard sexual cycle have been of tremendous value in extending genetic analysis to systems in which recombination is not based on meiotic crossing over. The results of these new adventures are evident in the successful mapping of bacteriophages and bacteria, and in the use of mitotic analysis in fungi (Chapter V). Furthermore, in bacteria, mapping has made use of a refreshing variety of natural and artificial processes: transformation, transduction, conjugation, and finally mechanical or radiation-induced fragmentation of the "chromosome."

The combined work of cytology and genetics in higher organisms over the last fifty years has led to a substantial understanding of the relations between linkage map and chromosome, and in particular between recombination fractions and incidence of crossing over at meiosis. It has also taught us to be aware of the limitations of genetic maps as pictures of the actual spacing of loci on the chromosomes. One has only to keep in mind how different the results are if we compare, for example, the loci in the X chromosome of *Drosophila melanogaster* mapped: 1) cytologically in the mitotic chromosomes by means of X-ray breakage; 2) cytologically in the salivary gland chromosomes; 3) genetically by means of meiotic crossing over, and 4) genetically by means of mitotic crossing over. This difference becomes embarrassing if we compare the meiotic maps of the male, in which crossing over does not occur, with those of the female.

In spite of these obvious limitations we must attempt to give a concrete meaning to the numerical relationships expressed in linkage maps. This means that we must aim at a description in chemical and physical terms of the chromosomes and of the processes taking place in them. Among these processes the most prominent are replication, crossing over, and the part played by the chromosomes in the metabolism of the cell. We are, of course, very far from this ambitious end. There are eminent geneticists who think that this is no concern of the geneticist but of the biochemist, the biophysicist, and the physiologist.

RESOLUTION AT THE INTER-GENIC LEVEL

As mentioned above, genetic analysis resolves pairs of loci not too far apart by detecting recombination between them and measuring its amount. The closer the two loci, the smaller the amount of recombination. Consequently the detection and measurement of recombination between two loci very close to each other requires the classification of a large number of products of meiosis: as long as we are below the ultimate limit—wherever it may be—the resolving power is determined only by the size of the sample that we are able, and are prepared, to classify.

Remarkable increases in resolving power have resulted in recent years precisely from increases in the size of the analysable samples. This is mainly the consequence of the introduction of microorganisms in genetical research and the development of selective techniques which pick out automatically rare recombinants from a mass of nonrecombinant cells. The result is that recombination fractions of one in a million have been measured. But larger samples can still be handled. There should be no insuperable technical difficulty in actually reaching the ultimate limit. Perhaps it has already been reached in the analysis of recombination in bacteriophage (Benzer, 1957) and Salmonella (Demerec et al., 1956).

Let us first consider the resolving power of genetic analysis using exclusively results from breeding experiments, as if we did not know of the existence and significance of chromosomes and as if we had no clue as to their chemistry. Table 1 gives examples taken from the literature of the closest linkages recorded between different genes in five organisms, ranging from a mammal to fungi. I am using here the

TABLE 1

EXAMPLES FROM FIVE ORGANISMS OF THE CLOSEST LINKAGES
RECORDED BETWEEN GENES PRESUMABLY NOT RELATED
IN PHYSIOLOGICAL ACTION

			Recombination	
Organism	_Total map units_	_Pairs of genes_	%	_as fraction $\times 10^4$ of total map_
DROSOPHILA	280	w and rst (I)	0.2	7
		sp and bl (II)	0.3	11
		ey and ci (IV)	0.2	7
MAIZE	904 } 1,350*	a_1 and sh_2	0.25	2
MOUSE	1,620 } 1,954*	se and d	0.16	0.8
ASPERGILLUS	660	y and $ad16$	0.05	0.8
		$ad15$ and $paba1$	0.5	9
NEUROSPORA	380 } 800*	$col4$ and $arg2$	0.4	5
		q and lys	0.5	6

* Indicates total maps calculated by multiplying the chiasma frequency by 50. Other total maps calculated by adding the lengths of all linkage groups. For those organisms for which both estimates are available, the larger has been used, as more likely to be nearer the true value. In the case of close linkages the recombination fractions, expressed as %, are equivalent to map units. (References: Mouse—Carter, 1955; Grüneberg, 1952; Slizynski, 1955; Aspergillus—Pritchard, 1955; Calef, 1957; Käfer, 1958; Neurospora—Singleton, 1953; Barratt, Newmeyer, Perkins, and Garnjobst, 1954; Maize—Darlington, 1934; Rhoades, 1950.)

term "gene" in the vague classical meaning based on the tacit and wrong assumption that the ultimate units of transmission and of difference in heredity are one and the same thing. This matter will come up for closer scrutiny in Chapter II.

In the examples given in Table 1, the two members of each pair of closely linked genes considered have quite different effects on the phenotype. It is reasonable to suppose that members of each pair are relatively independent in the part they play in metabolism or development. Their recombination fractions vary from 5 in 10^4 to 5 in 10^3. These recombination fractions represent from about 1 to 10 ten-thousandths of the total map for each organism.

We could follow Muller (1916, 1926) and use these ratios to make a minimal estimate of the total number of genes in each organism. It would be minimal, of course, because on the one hand there is no reason why the closest linkage found so far in any one organism should be just the closest occurring in that organism. On the other hand, as the detailed genetic analysis of an organism proceeds, the total identified map increases. In Drosophila, however, this analysis is so advanced that the map has not increased in the last thirty years or more. This means that the outermost genes of each linkage group already identified are really very near the ends.

RESOLUTION AT THE INTRA-GENIC LEVEL

When Muller made his estimates the following facts were not known. A gene, defined as a unit of physiological action (the matter is discussed later in this chapter and more fully in Chapter II) may have its chromosomal basis on a section of chromosome of considerable length and containing a number of "sites" of mutation (Pontecorvo, 1952), each separable from the others by recombination.

The work by Green, E. B. Lewis and Mackendrick with Drosophila, that by Roper, Pritchard, Forbes, Calef and others with Aspergillus, that by Giles with Neurospora and, foremost, the remarkable work by Benzer with phage T4 and by Demerec and co-workers with Salmonella, all point the same way. Recombination (be it by means of reciprocal exchanges as in crossing over or by other processes) resolves not only genes but also mutational sites of one and the same gene.

We have at present no evidence that the processes which recombine two sites within a gene are necessarily different from those which recombine two sites, one in each of two adjacent genes (see Chapter IV). In one case (Pritchard, 1955) there is an example of the recombination fraction between two sites in one gene being greater than that between one of these sites and another in a different gene nearby. The fact that we do not have evidence to the contrary does not exclude, of course, that there may actually be a difference between inter- and intra-genic recombination; a problem well worthy of a considerable research effort at the present moment (see Chapter IV).

In the majority of the cases analysed in Aspergillus and in Drosophila recombination between alleles of one gene does occur as a rare but regular event, not as an exception. The two complementary expected types of recombinants do arise and in some cases these complementary types have been recovered from one cell, i.e., they originate as a consequence of one recombinational event. It seems legitimate, therefore, to consider these cases as examples of crossing over between alleles. However, there are features of this intra-genic crossing over, and in general of crossing over within small intervals, which are not yet clear (see Chapter IV).

The examples of unidirectional transfer in heterozygotes described by Lindegren (1953)—who called them "conversion"—and Roman (1956) in yeast, by M. Mitchell (1955) and Case and Giles (1957) in Neurospora, by Strickland (1958) in Aspergillus, by Demerec (1928) in *Drosophila virilis,* and perhaps some of those described by Laughnan (1955) in maize, are very interesting though still completely obscure. They are unquestionably a source of confusion in certain cases of analysis of intra-genic crossing over. However, they are examples of something different from the intra-genic crossing over analysed in at least six allelic series in Aspergillus, and at least four in Drosophila. They are also different from the intra-genic recombination analysed by Benzer (1955) and Streisinger and Franklin (1956) in phage, by Morse, Lederberg and Lederberg (1956) in *Escherichia coli*, and by Demerec and co-workers (1956) in Salmonella. The difference is clear even on a purely formal analysis (Chapter IV). It would become even clearer if the results by H. K. Mitchell (1957) of differential effects on crossing over and conversion by temperature shocks were confirmed and extended.

We have now to consider the resolving power of recombination when it takes place between mutational sites within one gene (i.e., between allelic mutants) rather than between mutational sites of different genes (i.e., between non-allelic mutants).

When we test for recombination two mutants originated by distinct mutational events and allelic with each other, we are of course trying to resolve very closely linked sites, so much so that up to a few years ago crossing over between alleles was not known to occur. In some cases presumably the linkage may even be complete (a point hard to prove): mutation will have actually recurred at the "same" site, or there may be a minute structural rearrangement which prevents recombination.

It is not surprising that the examples of closest linkage so far detected come from tests between alleles. Table 2 gives examples of recombination fractions measured between alleles in various organisms. The smallest recombination fractions so far measured between alleles are of the order of 10^{-6} in Aspergillus and of 10^{-5} in Drosophila.

We can now attempt a minimal estimate of the total number of mutational sites in one organism from the ratio between the total map length and that of the smallest recombination fraction measured, following in this the argument used by Muller (1916, 1926) for estimating the total number of genes. There is here at least one assumption which must be made explicit. It is that the incidence of recombination between two adjacent mutational sites is the same along the whole map. The well-known fact that the distribution of

TABLE 2

EXAMPLES FROM FOUR ORGANISMS OF THE RATES OF
RECOMBINATION BETWEEN MUTATIONAL SITES OF
ONE GENE (CISTRON)

Organism, gene and reference	Mutational sites so far identified No.	Sum of recombination fractions between the two outermost sites (a)	Recombination fraction between the two closest (b)	Inferred minimal number of sites per cistron Ratios a/b	a/b*
DROSOPHILA					
1. bx	5	3×10^{-4}	3×10^{-5}	10	37
2. lz	3	1.4×10^{-3}	6×10^{-4}	2.5	187
3. w	4	5.6×10^{-4}	$8 \times 10^{-6*}$	70	70
ASPERGILLUS					
1. bi	3	1×10^{-3}	4×10^{-4}	2.5	1000
2. ad8	6	1.8×10^{-3}	15×10^{-6}	360	1800
3. paba	2		$1 \times 10^{-6*}$		
4. pro3	2		$1 \times 10^{-6*}$		
PHAGE T4					
1. r(II)A	39	4.3×10^{-2}	$1.3 \times 10^{-4*}$	330	330
2. r(II)B	18	3.5×10^{-2}	1×10^{-3}	35	269
3. h	6	2.0×10^{-2}	2×10^{-4}	100	153
SCHIZOSACCHAROMYCES POMBE					
1. ad2	3	4×10^{-4}	1.5×10^{-4}	2.5	65
2. ad7	9	1.5×10^{-3}	$6 \times 10^{-6*}$	251	251

* Indicates the smallest recombination fraction so far measured in each organism.
References: Drosophila—1. Lewis, 1954; 2. Green and Green, 1949, 1956; 3. Mackendrick, 1953 and unpublished; Aspergillus—1 and 3. Roper, 1950 and unpublished; 2. Pritchard, 1955; 4. Forbes, 1956; Phage T4—1 and 2. Benzer, 1955, 1957; 3. Streisinger and Franklin, 1956; Schizosaccharomyces Pombe—1 and 2. Leupold, 1937.
The data of Demerec and co-workers on Salmonella do not lend themselves to the treatment used in this table. They are very relevant, however, for the numbers of sites already identified as separable by recombination in transduction, e.g., 11 sites in the hiA gene (Hartmann, 1957); 12 sites in the cysB gene (Clowes, quoted by Demerec, 1956). The same can be said of the data of Roman (1956) with yeast, who has identified from 4 to 26 alleles as different in each of seven loci; out of a total of 83 independently arisen mutants at these loci, 73 were different as shown by the tests of complementary repair (see Chapter IV).

crossing over per unit of physical length of the chromosome is demonstrably not uniform does not necessarily make nonsense of this assumption. The chromosome may be, in fact, it probably is (Ris, 1957), differently packed along its length, so that the lengths measured under the microscope (e.g., Rhoades, 1950; Gall, 1956) may not have a uniform relation to lengths of the ultimate fibre.

The argument is as follows: on the assumption just made, the minimum recombination fraction so far measured in an organism can only be equal to, or an integer multiple of, the ultimate fraction, i.e., that occurring between two adjacent mutational sites. Clearly this estimate of the total number of sites—if the method is valid at all—can only be in error by defect: the smallest fraction so far measured is not necessarily the smallest occurring, and the map so far measured can only be smaller than or equal to the real total map. The results for three organisms are given in Table 3.

TABLE 3

MINIMAL ESTIMATES OF THE TOTAL NUMBER OF
MUTATIONAL SITES IN THREE ORGANISMS

	Total map units	Smallest recombination fraction measured		Total number of sites
			%	
	(a)		(b)	(a/b)
DROSOPHILA	280	w^{aE}-w'	8×10^{-4}	3.5×10^5
ASPERGILLUS	660	paba1-paba6	1×10^{-4}	6.6×10^6
PHAGE T4	800	rII55-rII247	1.3×10^{-2}	6×10^4

References: Drosophila—Mackendrick, 1953 and unpublished; Aspergillus—Roper, unpublished; Phage T4—Benzer, 1957, p. 91. The total map in phage is an extrapolation of the estimates from detailed mapping of a small region to the whole genome.

The conclusion is that the total number of mutational sites may well be of the order of ten million in organisms like Aspergillus and Drosophila and of ten thousand in organisms like bacteriophage (Table 3).

By the same sort of reasoning we can now go further and estimate the number of sites making up a gene. But before doing this we shall have to introduce some terminology, anticipating its full discussion in Chapter II.

In place of the term "gene," and to avoid its vague meaning, in previous publications (Pontecorvo, 1955; Pontecorvo and Roper, 1956; Pontecorvo, 1956) the terms "region," "section of allelism," or "set" have been used. They were meant for a portion of the linkage map, and, by extrapolation, of the chromosome, containing the mutational sites at which recessive mutants are allelic with one another. In its turn "allelism" was considered to be a purely functional relation, the operation for identifying which is perfectly clear (Pontecorvo, 1952, 1955): two recessive mutants *m1* and *m2* are allelic to one another when they are not complementary, i.e., when the heterozygote *m1/m2* has a mutant phenotype. Non-complementarity (which denotes unity of function) usually goes with the *cis-trans* (or Lewis) effect, i.e., the double heterozygote in *trans* is mutant but that in *cis* is not.

Benzer (1957) has adopted these two functional criteria and has proposed the term "cistron" for the map segment underlying a unitary function, as shown by non-complementarity and *cis-trans* effect of a set of recessive mutants. This very acceptable term will be used here instead of the term gene in every case in which it is less equivocal. It should be stated emphatically, however, that the criteria of allelism (and therefore those for defining a "cistron") as given above are not absolute. There are relationships between recessive mutants which are intermediate between allelism and complementarity. In addition, of course, those criteria cannot operate in the case of dominants (see Chapters II and III).

The number of sites in a gene or cistron can be estimated the same way as the total number of sites in the genome, i.e., dividing the total "length" of a cistron by the smallest interval measured, the assumption being again that the amount of recombination in any measured interval is either equal to or an integer multiple of an ultimate unit of recombination which, in its turn, is that between two adjacent mutational sites.

In cistrons in which more than two mutational sites have already been identified and located we can obtain this estimate by dividing the measured "length" of the cistron by the smallest element in that cistron. This would give an even greater underestimate of the total number of sites in that cistron than if we divided that "length" by the smallest element measured in the whole genome (marked with an asterisk in Table 2). Even this very conservative estimate gives figures of tens or hundreds of mutational sites per cistron. The less conservative estimate (last column in Table 2) gives figures reaching even into the thousands. In both these ways of estimating the number of sites per cistron, both the identified part of the cistron and the smallest segment within it are measured in a way which can only produce an error by defect: we are testing for recombination a small sample of mutants of independent origin of each cistron, and the probability is small that we have picked up in this sample both the two outermost sites of that cistron and two which are the closest together possible. For this reason the estimates are interesting only in the cases of high values: 360 sites in the *ad8* cistron of Aspergillus, 70 in the *w* cistron of Drosophila, 330 in the *rIIA* cistron of phage T4, and 251 in the *ad7* cistron in *Schizosaccharomyces pombe*.

How representative are these estimates for all the cistrons (genes) in the genome? Apart from the small number of different cistrons on which our conclusions are based, there is one important fact which suggests caution. The cistrons analysed for recombination between alleles of independent mutational origin are, obviously, cistrons of which a number of mutants were available. It is conceivable, to say the least, that cistrons with many mutational sites are precisely those more likely to yield mutants: our sample could very well be strongly biassed in favour of cistrons with an unusually high number of sites. As I suggested some years ago (Pontecorvo, 1952) we must keep an open mind and be prepared to find genes with all degrees of complexity, from those based on thousands of mutational sites to those based on few.

For this reason it does not seem very profitable at present to go one step further with speculation and calculate, from the total number of sites and the mean number in the two or three cistrons so far measured in three organisms, the total number of cistrons in an organism. If we wished to indulge in this, the results would be: 3,500 in Drosophila; 5,000 in Aspergillus, and 100 in phage.

But the important conclusion that at least some genes may have many mutational sites is supported from another angle, i.e., the proportion of independently arisen mutants which represent recurrence of mutation at the same site of a cistron, or in other words, the proportion of allelic mutants which, with a given resolving power, have not been found to be distinguishable from one another by means of recombination. Here the conclusion is wholly operational. In Drosophila, in which the classification of 10^4 gametes is already hard work, Green (1955a) and Green and Green (1956) for the *lz* and *f* cistrons, and Mackendrick (1953) for the *w* cistron, have failed to obtain recombination between quite a number of pairs of alleles. So has Dunn (1956) in the case of the *t* system in the mouse, and so have all the workers on human and animal antigens. But in Aspergillus experiments with a resolving power of 10^{-6} have so far never failed to yield recombinants between any two allelic mutants of independent origin. This statement is based on 23 mutants belonging to 7 cistrons:

Cistron	Alleles tested No.
ad8	6
bi	3
ad9	6
pro1	2
pro3	2
paba	2
Acr	2
Total	23

Dr. H. Levene has kindly calculated for me that the probability of obtaining by chance a distribution like this would be less than one in a million if there were no more sites in each cistron than those already identified; it would become about one in fifty for seven sites per cistron, and about one in four for fifteen sites per cistron, on the basis of equal probability of mutation per site and an equal number of sites per cistron.

If the mutability varied greatly from one site to another (as it certainly does) the results obtained would indicate an even greater number of sites. In fact, the finding, in Aspergillus, that out of nine intervals measured, three are extremely small (Table 2) shows that there are micro-regions of very high mutability: the "hot spots" of Benzer (1957). The work of Benzer (1957) and Streisinger and Franklin (1956) with phage, and of Z. Hartman (1956) and P. E. Hartman (1956) with Salmonella, and of Giles (1956) with Neurospora suggests that different sites of one cistron do differ grossly in mutation rates and that sites with high mutability may be clustered.

If the number of sites varied greatly from one cistron to another, again the results obtained would underestimate the number of sites. As to the extent of this variation in number of sites between cistrons, we have no clue.

Another way of estimating the number of sites per cistron could be that of comparing the average "distance" between two sites taken at random in a cistron with the average "distance" of two sites taken at random one in each of two adjacent cistrons. The greater the number of sites within a cistron, the greater the ratio of the average distance between two sites in one cistron to the average distance of two sites one in each of two adjacent cistrons. Unfortunately the extreme values of these ratios (for the calculation of which I am again indebted to Dr. H. Levene) are 2 in the case of only two sites per cistron and 3 in the case of an infinite number, the latter value being already attained for about twenty sites. This means that an analysis of this kind is not very sensitive and could give some information only if we had many more examples of analysis of several sites in a cistron and for many more adjacent cistrons.

Going back to Table 1, we have seen that most of the recombination fractions between neighbouring cistrons so far measured are of the order of 10^{-3}: though that between m and dy in Drosophila (Table 8) is as low as 10^{-5} (Slatis and Willermet, 1954). With one exception in Aspergillus (y to $ad16$: 0.0005; $ad16$ to $ad8$: 0.0014, Pritchard, 1955, see map on Table 10) these recombination fractions are larger than those measured between sites of one cistron, and this by a factor of the order of 10. The argument just developed suggests that the maximum ratio between average inter-cistron recombination and average intra-cistron recombination should be 3, for the case of very many sites per cistron, and provided that recombination between the two contiguous sites one of each of two adjacent cistrons were of the same kind and of the same frequency as that between any two contiguous sites of one cistron. The discrepancy between the expected factor of 3 and the found factor of 10 is obviously not significant for a number of reasons not only statistical, but it will be wise to keep it in mind.

The conclusions to be drawn so far are that genetic analysis has already resolved the linkage maps down to fractions of the order of a ten-millionth of the map. The results of measurements of this kind and of the location of mutants of independent origin suggest that the total number of mutational sites separable by recombination within a section of chromosome in which recessive mutants are allelic to each other (a cistron or gene, for short) may well be in the order of hundreds or thousands. Let us see whether we can give structural, or even better, chemical meaning to these conclusions.

RESOLUTION AT THE MOLECULAR LEVEL

This discussion so far has been based exclusively on the results of breeding experiments: the resolving power has been expressed exclusively in terms of recombination fractions as fractions of the total map, also based on recombination. Six years ago, repeating the pioneer attempt by Muller (1935), I tried to translate the meagre measurements of minute recombination fractions available then into lengths of chromonema. This required using recombination fractions from one organism (Aspergillus) and length of the presumably fully stretched chromonema from another one (Drosophila, salivary gland chromosomes). We can do better now that Benzer's (1955, 1957) brilliant analysis of the *rIIA* cistron in bacteriophage T4 has shown what can be done. For a few organisms besides phage, we know the DNA content of the haploid complement or particle, and we can express the smallest recombination fractions measured as fractions of this DNA content (Pontecorvo and Roper, 1956).

The underlying assumptions are disputable even in the case of phage. They are rash in the other cases. We attempt these calculations only because they may be useful in bringing to light what the problems are, and in thinking out experiments. The assumptions are (Benzer, 1955; Pontecorvo and Roper, 1956): 1) that DNA constitutes the structural backbone of the linkage map and that all the DNA is in this backbone; 2) that recombination involves exchanges of nucleotide linkages either by a process of breakage and reunion or by one of change in copy-choice (Lederberg, 1955); 3) that each nucleotide linkage, or small multiple of nucleotide linkages, has the same probability of taking part in recombination as any other one; and 4) that mutational sites are separated from one another, along the whole length of the chromosome, by one nucleotide linkage (better, a pair in the Watson-Crick model) or by a fixed small multiple of pairs of nucleotide linkages.

In the case of bacteriophage T4, on which Benzer's analysis was based, there is at least some evidence that the first assumption is approximately true, though Levinthal's (1956) experiments suggest that not all the DNA qualifies. If the genetic information in phage is contained in the DNA, it is also possible that nucleotide linkages are actually those involved in recombination.

The results of Table 4 are for five organisms, for which measurements of total linkage map, of total DNA per genome and of very small recombination fractions are available. They confirm Benzer's suggestion for phage by extending it to other organisms on the assumptions made above. The smallest recombination fractions measured to date would, for at least some organisms, represent only a few nucleotide pairs. This appears to be the case not only in phage, but also in *Escherichia coli* and *Aspergillus nidulans* and perhaps Drosophila, and it is conceivable that it could be so also for the mouse and maize if only the resolving power of recombination were increased sufficiently.

The next step in this speculation, still following Benzer, is that the ultimate unit of crossing over is a single nucleotide bond (3.4 A.)

TABLE 4

SMALLEST RECOMBINATION FRACTIONS MEASURED IN SIX
ORGANISMS EXPRESSED AS FRACTIONS OF THE TOTAL DNA

	Total linkage map units	Total DNA in nucleotide pairs	Nucleotide pairs per map unit	Minimum recombi- nation fraction measured (a)	(a) as fractions of total map	(a) as fraction of total DNA in nucleotide pairs
1. PHAGE T4	800	2×10^5	2.5×10^2	1×10^{-4}	1×10^{-5}	2
2. ESCHERICHIA COLI	2,000	1×10^7	5×10^3	2×10^{-5}	1×10^{-6}	10
3. ASPERGILLUS	660	4×10^7	7×10^4	1×10^{-6}	1.5×10^{-7}	3
4. DROSOPHILA	280	8×10^7	3×10^5	8×10^{-6}	2×10^{-6}	40
5. MOUSE	1,954	5×10^9	3×10^6	2×10^{-3}	1×10^{-4}	500.000
6. MAIZE	1,350	7×10^9	8×10^6	2×10^{-3}	2×10^{-4}	350.000

References: Phage T4—Benzer, 1957; *Escherichia coli*—Jacob and Wollman, 1958; Aspergillus—Pontecorvo and Roper, 1956, brought up to date; Drosophila—Kurnick and Herskowitz, 1952; Mackendrick, unpublished; Mouse—Vendreli, 1955; Grüneberg, 1952; Maize—Rhoades, 1950; Ogur and Rosen, 1950.

or perhaps a fixed multiple of nucleotides, say ten, i.e., one complete turn in the Watson-Crick double helix or 34 A. A further speculation is that a mutational site is the segment of duplex DNA helix between two nucleotide linkages relevant for crossing over, i.e., one nucleotide pair or a small number of nucleotide pairs, say ten. A cistron or gene would then work out to be based on a length of helix, say, 1,000 nucleotide pairs long. Perhaps this is what is needed to determine the specific arrangement of aminoacids in a protein (cf. Ingram, 1957; Crick, 1958). In terms of only one Watson-Crick duplex helix as the basis of the linkage map, the length of a cistron would work out as some 3,400 A. and correspondingly less if the basis of the linkage map were a multi-stranded structure made up of several Watson-Crick helices.

The trouble in all this—apart from the unwarranted assumptions on which it is based—is that our analysis measures the internodes between mutational sites and we are now identifying the separators with the separated. Even in phage, where it looks as if nothing but the DNA carried the genetic information, this is a rather hazardous step. In higher organisms it is completely gratuitous. Let us examine some of the difficulties to which it leads.

In closely related groups, like the Anura and the Urodeles, the DNA content of the nucleus differs by a factor of 20 times: does this mean that there are 20 times more genes, or that genes are 20 times more complex, or that the bulk of the DNA has nothing to do with linkages involved in crossing over, or that the multiplicity of DNA helices of the chromonema can vary as much as from one to twenty between two groups of organisms even closely related taxonomically?

Another point: Table 4 shows that the amount of DNA per unit of recombination increases by a factor of 10,000 times when we consider a series of organisms starting with phage and ending with the mouse or maize. This matter will come up again in Chapter IV. It certainly shows that the relations between recombination and DNA are not direct and simple.

Clearly the attempt to give chemical content to the resolving power of genetic analysis is useful in the case of bacteriophage but perhaps not yet in the case of higher organisms. We do not know, in chemical terms, what we are trying to resolve and into which elements. Yet, in purely genetical terms we can express the resolving power of genetic analysis quite validly: this resolving power is at present of the order of one ten-millionth of the linkage map. Even if the whole material of the nucleus went up to make nothing but the linear bonds which we resolve, this resolving power would still be astonishing.

Genetic techniques turn out to be more sensitive than biochemical ones, as I contended at the beginning of this chapter. They make profitable use of the enormous amplifying system of the organism. Biochemistry, however, is beginning to catch up and attempt to give structure, i.e., spatial organization, to chemical reactions within the cell. Dr. Hotchkiss, my predecessor in the Jesup Lectures, has shown that genetic analysis of transforming principles of Pneumococcus, which are pure DNA, detects by means of recombination differences between two parts of a macro-molecule. Biochemistry cannot yet do so when, as in the case of transforming principles and genes, that macro-molecule is one out of thousands of not very different ones, or better, is a part of an enormously greater aperiodic complex (Schrödinger, 1943). Surely there is a case for joining forces in the field of the fine structure of the genetic materials.

The present chapter contains a number of statements of faith and day-dreams rather than documented inductions or deductions. The documentation is reserved for later chapters, but some of the statements of faith will remain such. Science would not be able to advance as it does if it had to rely exclusively on the scientific method.

CHAPTER II

ALLELISM

THE ambitious theme of this chapter is no less than to discuss what the ultimate units of heredity are. This is, of course, a theme which has been in the foreground of theoretical discussions from the beginning of genetics. Its clear formulation, however, is not so old: it was put into a nutshell by Muller about twenty years ago (Raffel and Muller, 1940) in a discussion on the divisibility of the genetic material.

I shall consider two theses. One is that the idea of a single entity —the gene—with the simultaneous properties of unit of hereditary transmission, unit of mutation, and unit of function, is no longer adequate. I do not imply that this idea has not been or is no longer useful: it is such a good approximation of the truth that it has made possible the whole development of genetics. It can still be used for advancing fundamental and applied genetics in practically all but one direction: it is past use as a working model when we wish to understand the fine organisation of the genetic material (Goldschmidt, 1955). The fallacy is analogous to that of considering a nation as an entity which occupies a particular territory, belongs to a particular race, and has a particular culture.

After discussing in what ways the three properties mentioned above are not coextensive, i.e., they are not properties of one and the same thing, the second thesis will be the following. The three properties themselves can be defined exactly (e.g., Benzer, 1957) but the examples found experimentally show a range of situations which make the definitions not applicable to all, or even most, cases.

It will be wise, therefore, to keep an open mind. The examples which we investigate, because they fit best the definitions, could represent only one type—the simplest—of organisation of the genetic material, and there may be others more complex.

RECOMBINATION AND FUNCTION

The gene was supposed to have three properties: indivisibility in heredity, specificity of function, and ability to mutate. More in detail the gene was supposed to be 1) an ultimate unit in inheritance, i.e., not further subdivisible by either recombination or chromosome breakage; 2) an ultimate unit of phenotypic difference, i.e., associated with a single primary specific function in metabolism or development, a function not further divisible, and 3) an ultimate unit of mutation, i.e., the smallest part of a chromosome which, when changed, would be replicated in the changed form.

"Allelism" was a special relation between two elements of the genetic material descending from the three properties just mentioned. Any two forms of a gene, differing from one another according to 3) were called "alleles" of one another. This implied, according to 1), that two alleles could not recombine; they had to segregate from one another at gametogenesis in a heterozygote and they could not be both in the same chromosome. It also implied, according to 2) that they would determine a difference in one primary function.

We shall see first what the difficulties are as to the presumed non-recombinability of alleles, and second as to the presumed specificity of functions.

Let us start with recombination. Consider two haploid strains of yeast, one requiring arginine for growth, and one, from which the former was derived by mutation, not requiring it. We combine the two into a diploid and the diploid does not require arginine; i.e., the requirement is recessive. We cross the two sexually and tetrad analysis shows a perfect 1:1 segregation, i.e., each of a large number of asci gives two ascospores which develop into strains not requiring arginine and two ascospores which develop into strains requiring it. In practise we would conclude that the two original strains differed at a single locus, the requiring strain having a recessive allele and the other a dominant allele at that locus. We would reach the same conclusion if, instead of analysing all products of each ascus, we classified a large number of strains originating from a random sample of ascospores. This is, of course, the procedure more generally used, in fact exclusively used, in higher animals and plants.

That conclusion could be incorrect, as an example will show. Consider a case in which it is known that one difference in phenotype is the result of interaction between two non-allelic recessives. Suppose we had a true breeding strain of Drosophila with white eyes, because homozygous for *cn* and *bw*, and we crossed it to the wild type, getting an F_1 all wild, and backcrossed the F_1 males. We would again get a 1:1 ratio of white to wild because of the absence of crossing over in the male. If, on the other hand, we backcrossed the F_1 females we would get, of course, the four types, red, white, cinnabar, and brown in about equal proportions. But if the *cn bw* strain had a suitable rearrangement preventing crossing over between the two markers, even the backcross of the females would give 1:1 ratios. Alternatively, if *cn* and *bw* were very closely linked, again the backcross of the females might give a 1:1 ratio.

There are operations, thus, which can exclude the possibility that a difference in phenotypes is determined by only one difference between two homologous chromosomes, but there is no operation that will tell us that it is determined by just one. In other words, we can have positive evidence of non-allelism, but not of allelism—if our criterion is that of recombination.

We could, of course, call "allelic" any localised difference in homologous chromosomes which did not recombine in a test of 10^6 gametes or of any other arbitrary number upon which an appropriate International Committee would agree.

Thus a first point is clear: the criterion of recombination does not discriminate between allelism and non-allelism. It would become an absolute criterion if the ultimate frequency of recombination were known and the experiments were designed to classify a number of gametes well in excess of what would be required by this ultimate frequency. A situation of this kind has perhaps already arisen in phage (Benzer, 1957) and Salmonella (Demerec, 1956).

There is another criterion which we can use: the functional test of complementarity. In Drosophila *cn* is recessive, and *bw* is recessive: a cross between homozygous *cn* and homozygous *bw* gives a double heterozygote which has the double dominant phenotype, i.e., red. This test can apply only to recessives. It becomes more illuminating if applied to recessives determining similar phenotypes.

For instance, let us take the example of the adenine requiring mutants of *Aspergillus nidulans* (Pontecorvo, 1956). Many strains of this type are available and the bulk of them are qualitatively indistinguishable from one another at the crude level of growth response to substances metabolically related to adenine. About fifty have been tested in crosses and each turned out to differ from the wild type at one locus as determined by a test of the type just described. The mutants fall into eight groups. The members of each group—with one interesting exception to be discussed presently—are complementary to those of the other seven but not so among themselves. Two groups are not yet located; two (*ad1* and *ad3*) are very closely linked (0.1 percent recombination); the other four are not. All six groups already located fall into two out of the eight (Käfer, 1958) linkage groups of *Aspergillus nidulans* (Table 5).

In this classification we find that, with the exception of *ad1* and *ad3* just mentioned, the location of the mutants agrees, in first approximation, with their subdivision into the six groups on the functional basis of complementarity. That is, a crude location of the mutants (by measuring the recombination frequency of each relative to other markers the locations of which are already known) gives five positions corresponding to the six complementary groups. We can then test for recombination, two by two, the mutants within each group, and we find that out of a sample of, say, a few hundred products of meiosis, there are no adenine independent recombinants. We would conclude in classical genetics that, with the exception of *ad1* and *ad3*, the mutants in each position are allelic because: 1) they are recessive, and in heteroallelic (Roman, 1956)

TABLE 5

LINKAGE MAPS OF *ASPERGILLUS NIDULANS*
(*as of November 1, 1957*)

CHROMOSOME I

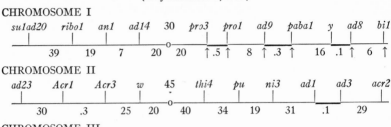

su1ad20   ribo1   an1   ad14 30 pro3 pro1 ad9 paba1 y ad8 bi1

39   19   7   20 20 ↑.5↑ 8 ↑.3↑ 16 .1↑ 6↑

CHROMOSOME II

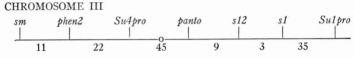

ad23 Acr1 Acr3 w 45 thi4 pu ni3 ad1 ad3 acr2

30 .3 25 20 40 34 19 31 .1 29

CHROMOSOME III

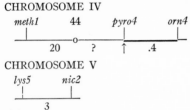

sm phen2 Su4pro panto s12 s1 Su1pro

11 22 45 9 3 35

CHROMOSOME IV

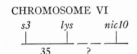

meth1 44 pyro4 orn4

20 ? ↑ .4

CHROMOSOME V

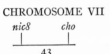

lys5 nic2

3

CHROMOSOME VI

s3 lys nic10

35 ?

CHROMOSOME VII

nic8 cho

43

CHROMOSOME VIII

co ribo2 cys2

40 ?

Mapped total genome: 660 units. Identified total loci: 41.

Figures between symbols of genes indicate recombination fraction % based on Käfer's (1958) summary. Cases of close linkage (<1 unit) are indicated with heavy line. Cistrons analysed in detail are indicated by arrows.

Symbols: o = centromere; *acr* or *Acr* = acriflavine resistance (recessive or semi-dominant); *ad* = adenineless; *an* = aneurinless; *bi* = biotinless; *cho* = cholineless; *co* = compact colony; *cys* = cystineless, *lys* = lysineless, *meth* = methioninineless; *ni* = unable to reduce nitrate; *nic* = nicotinicless; *orn* = ornithineless; *paba* = p-amino-benzoicless; *panto* = pantothenicless; *phen* = phenylalanineless; *pro* = prolineless; *pu* = putrescineless; *pyro* = pyridoxinless; *ribo* = riboflavineless; *s* = unable to reduce sulphate; *sm* = small colony; *Su* or *su* = suppressor dominant or recessive, followed by symbol of suppressed gene; *thi* = "thiazole"-less; *y* = yellow conidia; *w* = white conidia (epistatic to *y*/+).

combinations in a diploid they determine the recessive phenotype; and 2) they are located in the chromosomes at the same position. In other words, within the resolving power of the sample, we are unable to detect any difference in position.

Yet, if we pushed the analysis further, this second conclusion would turn out to be premature. If instead of 500 we test 50,000 or 500,000 products of meiosis, we would find among them, from combinations of two mutants of the same group, the two classes of recombinants: wild type and double mutant. I shall explain in Chapter IV how wild type and double mutant are recognised and how it is shown that they are indeed *the* two reciprocal products of recombination.

Thus the functional criterion of non-complementarity, i.e., that the diploid heteroallelic for two recessive adenine-requiring mutants is adenine-requiring, tells us that the two mutants are allelic, but the recombination criterion tells us that they are not.

One important fact, however, is clear: functional allelism (or non-complementarity) goes with approximately identical location, or, in other words, very close linkage, i.e., usually less than one in a thousand recombination. But the reverse is not necessarily true as shown by *ad1* and *ad3*: these two recessives are located very close together, again with recombination of the order of one per thousand, and yet they are fully complementary. The compound *ad1* $+/+$ *ad3* has the dominant phenotype (see map, Table 5).

Pairs of recessives having presumably unrelated effects and showing complementarity and close linkage (with recombination of the order of one per thousand) are quite common. Table 1 gave examples of this kind of relation. In the chromosome segment studied by Pritchard (1955) in Aspergillus *y* and *ad16*, two complementary mutants—one affecting the colour of the conidia, the other determining a nutritional requirement—are closer than *ad16* and *ad8*, two mutants which are functionally allelic, i.e., noncomplementary.

Thus complementarity is compatible with close proximity down to the point of no resolution by recombination. Similarity of effects, down to the point of apparent identity (Fincham and Pateman,

1957), is also compatible with complementarity. This is clearly shown by the extensive work on: the *t* series in the mouse (Dunn, 1954); *rIIA* and *rIIB* in phage (Benzer, 1955); *ad1* and *ad3* (Pontecorvo, 1952), *pro1* and *pro3* (Forbes, 1956), and *ad15* and *ad17* (Calef, 1957) in Aspergillus; *pdx* and *pdxp* (Mitchell, 1956), am^{32} and am^{47} (Fincham and Pateman, 1957), B3 and B5 (Case and Giles, 1957) in Neurospora; and numerous examples in Drosophila, Salmonella, etc. (See Table 8, Chapter III).

In Drosophila, particularly notable is the example of *m* and *dy* where the recombination fraction of 5×10^{-5} (Slatis and Willermet, 1954) is the lowest (other than 0) measured between complementary genes. On the other hand, the highest recombination (0.5 percent) so far measured between two noncomplementary (i.e., functionally allelic) recessives in organisms higher than phage is that found by Koske and Maynard-Smith (1954) between two *ar* alleles of *Drosophila subobscura*. I am indebted to Professor J. B. S. Haldane for calling my attention to this case.

We come to the conclusion then, that while complementarity of pairs of recessive mutants (with related or with unrelated effects) may or may not go with close linkage, non-complementarity always goes with it. As far as it is known, non-complementarity also always goes with related and often indistinguishable effects, as exemplified by the *ad* alleles of any one of the six groups mentioned before.

Summarising, pairs of recessive mutants are known showing any one of the following relations:

	Effects	*Recombination other than 0*
Complementarity	similar or not	down to 5×10^{-5}
Non-complementarity	similar	up to 5×10^{-3}

These relationships could be taken to suggest (Pontecorvo, 1955) that the only objective criterion for classifying as allelic or non-allelic closely linked mutants with similar effects is that of non-complementarity or complementarity, i.e., phenotypic and functional.

But here we are faced with an additional fact. The distinction between complementary and noncomplementary is not an absolute one even for the case of recessives; and, of course, is meaningless in the case of dominants.

For example, in the *ad9* series of adenineless mutants of Aspergillus, Calef (1957) and Martin-Smith (1958) have found the following situation. Six independently arisen mutants—all recessive to the wild type both in the heterozygote and heterokaryon, all "isoalleles," i.e., phenotypically indistinguishable—are located 0.3 units from another marker. Mitotic recombination (see Chapter V) between the *ad9* mutants occurs in all the fifteen diploids heteroallelic for two-by-two combinations of these six mutants and gives this tolerably linear order:

$$ad33 - \genfrac{}{}{0pt}{}{ad13}{ad9} - ad17 - ad15 - ad32.$$

(The position of *ad9* relative to *ad13* is not yet known.)

The peculiarity of this case is that two out of the fifteen heteroallelic combinations in diploids are inconsistent with a simple system of functional allelism of all six mutants. Precisely, *ad17* is semicomplementary to *ad15* (Calef, 1955) and fully complementary to *ad32* (Martin-Smith, 1958): i.e., the heteroallelic diploids *ad17/ad15* and *ad17/ad32* are near to or indistinguishable from the wild type, while all the other thirteen heteroallelic diploids, including combinations with *ad17*, are adenine-requiring. It is interesting to note that the two aberrant relations involve three sites which at the present stage of analysis are contiguous and terminal. The situation is different from that found by Green and Green (1956) in the lozenge series in Drosophila. Here two mutants (*50e* and *49h*) are complementary to all the fourteen others: there are no pairs of inconsistent relations.

Clearly, even the functional criterion of complementarity does not permit us to draw an absolute distinction between allelism and nonallelism. If we consider examples of inheritance with manifestation of both alleles in a heterozygote, like those which are the rule in the inheritance of antigens, this conclusion is even more obvious.

In the case of alleles with manifestation in the heterozygote, controversies such as those of whether or not the various *Rh* types are examples of multiple allelism or close linkage have always been operationally meaningless. We realise now that they would be meaningless even if the supposedly critical operation—a test of recombination—were feasible. If recombination between, say, *D* and *C* were actually observed, this would no more help to decide for or against the label of multiple allelism than the many analogous cases in Drosophila, Aspergillus, phage, and Salmonella, in which it has been observed (cf. Whiting, 1950). What is vague is, of course, the concept of allelism itself. Also to be noted is the following fact. In the case of *Rh* the sensitivity of the phenotypic comparison (serological) is very great, but that of the recombination test is practically nil. In the other examples in Drosophila, Aspergillus, phage and Salmonella, the sensitivity of the phenotypic comparison (visible difference or difference in growth response) is very crude, but the recombination test very sensitive.

From all this semantic confusion something constructive emerges; there are all possible intermediate situations between two extremes. At the one end we have two recessives located in two minute segments wide apart on the same chromosome pair or in two different chromosome pairs, and determining apparently unrelated phenotypic effects; the dominant allele in one chromosome complements the recessive in the homologue and vice versa. Examples of this in classic genetics are any pair of genes unlinked or loosely linked affecting different "characters." At the other end we have two recessives within one minute segment at homologous positions on each member of a chromosome pair. The mutants show noncomplementary and phenotypically indistinguishable effects; recombination between them may be obtained, but if so usually only with a high resolving power. We could use the term "allelism" for functional relations of this second kind with the understanding that we are dealing with one extreme of a distribution.

Demerec (1956) has proposed the useful term "nonidentical alleles" for pairs between which recombination has been obtained.

The nonidentity refers to the fact that each represents mutation at a different site, as shown by recombination. Another useful term is "heteroallelic," proposed by Roman (1956) for combinations of two alleles of different mutational origin which yield the wild type by recombination or other mechanisms.

THE CISTRON

Most of the cases of allelism, in the sense now defined, which have been studied thoroughly—and they number about twenty in organisms ranging from phage and *Escherichia coli* to Drosophila, Aspergillus and the silkworm—show what E. B. Lewis (1951) calls "position pseudoallelism" and what I prefer to call noncommittally the "Lewis effect" (Pontecorvo, 1955). That is, a heterozygote having two different allelic recessives (arisen by independent mutations), one on one chromosome and the other on the homologous chromosome (*trans* or *repulsion* arrangement), has a recessive phenotype, or a more nearly recessive phenotype than the corresponding double heterozygote with both recessives on one chromosome and a normal homologue (*cis* or *coupling* arrangement):

$\dfrac{m_1 \ +}{+ \ m_2}$ has a more nearly recessive phenotype than $\dfrac{m_1 \ m_2}{+ \ +}$.

Incidentally, the terms *cis* and *trans* which are now in common usage, were introduced into genetics by Haldane (1942) in place of the Batesonian "coupling" and "repulsion" in crosses involving linked segregations. They were later applied by me (Pontecorvo, 1950) to problems of the kind under discussion.

The *cis-trans* effect, or "position pseudoallelism," or "Lewis effect," clearly indicates a strictly localised functional integration between two or more closely linked elements of the genetic material.

That this is not merely a matter of proximity but of some special type of functional integration is shown in examples like the one already mentioned of the relations between *y* and the mutant alleles of the *ad8* series in Aspergillus (Pritchard, 1955, 1958). The recessive *y* determines a change in colour of the conidia from wild-type green to yellow. The mutants of the *ad8* series determine require-

TABLE 6

EXAMPLES OF ALLELIC SERIES IN WHICH THE *CIS-TRANS* EFFECT HAS BEEN LOOKED FOR AND FOUND

Organism, cistron and reference	No. of sites so far resolved
DROSOPHILA	
Star (Lewis, 1945)	2
Stubble (Lewis, 1951)	2
bithorax (Lewis, 1954)	5
lozenge (Green and Green, 1949)	3
white (Mackendrick and Pontecorvo, 1952; Lewis, 1952; Judd, 1957)	4
forked (Green, 1955a)	2
vermilion (Green, 1954)	2
singed (Ives and Noyes, 1951; Hexter, 1955)	3
garnet (Chovnick, 1957)	3
dumpy (Carlson, 1957)	3
SILKWORM	
E (Tsujita and Sakaguchi, 1953)	2
ASPERGILLUS	
ad8 (Pritchard, 1955)	6
ad9 (Calef, 1957; Martin-Smith, 1958)	6
bi (Roper, 1950)	3
paba (Roper, unpublished)	2
ESCHERICHIA COLI	
Gal (Morse, Lederberg and Lederberg, 1956)	2
Lac (E. Lederberg, 1952; J. Lederberg, 1951)	2
PHAGE T4	
rIIA (Benzer, 1957)	39
rIIB (Benzer, 1957)	18
h (Streisinger and Franklin, 1956)	6

ment for adenine; y is, of course, complementary to all alleles of the *ad8* series. The recombination between y and one of the nearest mutants (*ad16*) of the said series is 0.05 ± 0.05 percent and that between the latter and another mutant of that series (e.g., *ad8*) is 0.14 ± 0.04 percent.

In other words, the "distance" between two genes, which are different on all criteria (recombination, complementarity, and gross phenotypic effects), is not greater, in fact, is almost certainly smaller, than that between two sites of one of them. Yet, the phenotype of the heterozygote $\dfrac{y \quad ad16}{+ \quad +}$ is the same (wild type) as that of the heterozygote $\dfrac{y \quad +}{+ \quad ad16}$, whereas those of the heterozygotes

$\dfrac{ad16 \quad +}{+ \quad ad8}$ and $\dfrac{ad16 \quad ad8}{+ \quad +}$ are quite different, the former being mutant

and the latter wild type (Roper and Pritchard, 1956).

Thus in the *cis-trans* test we have an additional criterion to that of non-complementarity for concluding that the degree of functional integration between the sites of one gene is more intimate than that between the sites of two genes. We have, in fact, an objective test, purely genetical and not requiring any biochemical analysis, for defining a unit of function in heredity. As far as it is known, a unit of function defined by the *cis-trans* test has usually a structural basis not overlapping that of its adjacent neighbour; but this is not general.

In what precisely the unity consists is a matter for speculation which will be discussed later. There is, however, considerable attraction in the hypothesis that it has to do with the synthesis of individual proteins: in other words, with some modified form of the "one gene-one enzyme" hypothesis of Beadle (1945).

As mentioned in Chapter I, to the ensemble of mutational sites at which mutants are noncomplementary to each other and show the *cis-trans* effect, Benzer (1957) has given the name of "cistron." This operational definition of a functional unit seems to be the only objective one possible at present. Even though already it does not fit all the known examples (see, e.g., the *ad9* series), it fits most of them well enough (Table 6), and it is likely to be useful.

Let us consider the alternatives, and the difficulties into which they lead. As one alternative, one could take identity at the crude level of phenotypic comparisons as denoting identity of impaired function between independently arisen mutants. This would lead to the absurdity, e.g., that genes operating on successive steps of developmental or biochemical reactions would be classed as having the "same" function, if our phenotypic comparison were at the level of the end-product. For instance, the eight different groups of adenineless mutants in Aspergillus would be classified as being impaired in the same function.

Even considering only the cases of close linkage between mutants of this kind, the difficulty remains. It leads to such contradictions as the following one (Green and Green, 1956, cf. pages 709 and 714): The recessive amx, which determines a phenotype similar to lz^{BS} (complementary and about 0.1 unit away), is considered "unrelated to the mutants of the lz pseudoallelic array and should not be included as part of the array." On the other hand, the recessive lz^{50e}, which again determines a phenotype similar to lz^{BS}, to which again it is complementary, but with which it has not yielded recombinants, is considered to be "truly allelic."

As a second alternative, one could beg the question of the one gene-one enzyme hypothesis and consider as involved in a single primary function two or more mutants which are impaired in the synthesis of the same protein. I am aware of only one example so far of thorough chemical analysis of a gene-induced change in a specific protein: sickle-cell anaemia in man (Ingram, 1957). There are in addition about ten examples (Horowitz, 1956) from Neurospora, and about as many from *Escherichia coli*, of mutational loss of activity of specific enzymes or change of characteristics in the activity of specific enzymes. Clearly, research of this kind is what is needed, provided it goes as far as that on the sickle-cell haemoglobin.

If it does not, we are always left in doubt. For instance, Fincham and Pateman (1957) have found that two Neurospora mutants (am^{32} and am^{47}) show no detectable glutamic dehydrogenase activity. Crosses involving the two have so far failed to yield unquestionable recombinants. The two mutants, however, are complementary. Clearly only the isolation and complete analysis of the enzyme could exclude the possibility that we are dealing with two proteins working as a unit. Until this is done (and it is a difficult requirement if we exact it as a routine) we cannot conclude anything about the primary function or functions impaired in the two mutants.

In conclusion, for the time being the only simple, objective, and workable definition for a gene-determined unit of function lies in the empirical test of non-complementarity and the correlated *cis-trans* effect. In applying this test we must remember again that it is probably not absolute.

What does non-complementarity of recessive mutants (and the usually associated *cis-trans* effect) imply? In a heterozygote $\frac{+}{m_1}$ the normal, dominant structure $(+)$ is capable, at least qualitatively, of carrying out the function in which the mutant, recessive structure (m_1) fails. In a double heterozygote between complementary mutants $\frac{+\ m_2}{m_1\ +}$, or $\frac{m_1\ m_2}{+\ +}$, the part of one structure, normal in one respect, is capable of carrying out the function which fails in the other and vice versa.

If m_1 and m_2 are noncomplementary it means that the normal part of the structure defective in m_1 and that of the structure defective in m_2 cannot compensate reciprocally for the impairments. However, when the arrangement is such that one structure is defective in both m_1 and m_2 and the other is completely normal, compensation takes place. We take it that this defines operationally "one function."

By analogy—suggested by certain remarks of Hotchkiss (1957)—if I have four cars, two Cadillacs and two Chevrolets, and one Cadillac and one Chevrolet have broken down, the other two can still function and tow the defective ones to the nearest garage. But if I have only two Cadillacs and the distributor of one and the carburetor of the other have broken down, transport will come to a standstill. It can continue if I make one good car from two defective ones by moving the good carburetor to the car with the good distributor. I can now tow the doubly defective Cadillac by means of the now functional one. The fact that the ability to tow is restored only by moving the sound parts from one Cadillac to another defines the two cars as functionally alike, i.e., Cadillacs, as distinct from two cars functionally different, i.e., a Cadillac and a Chevrolet.

Going back to the mutants m_1 and m_2 we could say that the difference between allelism and non-allelism arises at the transition between complementarity only in the *cis* arrangement, and complementarity irrespective of arrangement.

This transition defines the "cistron" in the case of recessive mutants. We shall discuss later how to fit into this scheme the case of genes with manifestation of both alleles in the heterozygote, e.g., genes determining antigenic differences, enzymatic differences (Horowitz and Fling, 1956), and other protein differences (Pauling, Itano, Singer and Wells, 1949; Ingram, 1957).

RECOMBINATION AND MUTATION

The mutational sites of one gene or cistron identified by recombination are likely to be many, as shown in Chapter I, though different cistrons may well differ in this respect. There are a number of questions relevant to allelism arising from this conclusion, and some will be considered now.

One concerns the relationships between mutational sites resolved by recombination and the ultimate units of mutation. As mentioned in Chapter I, the way in which we resolve mutational sites tells us nothing of the sites themselves, but only something of how they are joined to each other: e.g., that they are in a linear order.

The conjectural picture of the backbone of the chromonema as one duplex Watson-Crick fibre of DNA would imply the possibility of loss or change in a single nucleotide pair or in a number of nucleotide pairs or change in the sequence of nucleotide pairs anywhere, with recombination also liable to occur at any nucleotide linkage. A mutational site could be, then, one nucleotide pair and a cistron would be based on anything up to 1,000 or more such nucleotide pairs separable by recombination and each capable of change, among other ways, perhaps *via* the tautomeric changes suggested by Watson and Crick.

A change of bases, or loss of one nucleotide pair or a rearrangement, including deletion, in the sequence of nucleotides, either minute (i.e., within the cistron) or gross (i.e., with one break within the cistron and one elsewhere) would constitute a mutation. The result, if still compatible with replication, would yield a mutant cistron, behaving as a functional allele of that from which it arose and of other mutants of that cistron.

To be more cautious and use instead the less committal picture of a linear series of mutational sites not better specified in chemical terms, we could say that a cistron may include at least tens of mutational sites, each capable of change, possibly in more ways than one, or of change in arrangement relative to other sites of the same as well as of other cistrons. Either type of change if replicable would be a mutation and "point" mutations would be, as we all know they are, indistinguishable in results from minute rearrangements.

There seems to be sufficient evidence in the case of phage and Salmonella (Benzer, 1957; Demerec, 1956), that one site can mutate in more ways than one from the wild type. On the other hand, Streisinger and Franklin's (1956) work, again on phage, suggests that a mutant at one site has usually only one way of mutating back to wild. This, however, may not be generalisable in view of the possibility of what we could call "intra-cistron" suppressors described further on.

The study of the mutational pattern of individual genes is linked pre-eminently to the name of Stadler (summary: Stadler, 1951). His great contribution, both in experiments and ideas, has had a powerful stimulus, not consciously realised by most research workers, in canalizing research towards the analysis of the fine organisation of individual genes.

The result is already most rewarding with microorganisms, especially phage, *Escherichia coli*, yeasts and Aspergillus, but perhaps less so with higher organisms, where the resolving power is much smaller. It is therefore inevitable, though regrettable, that an analysis of the mutation pattern as refined as that at the A and R loci in maize (Stadler, 1951) cannot be linked up with an equally refined analysis of the recombination pattern. The same applies to D. Lewis's (review: 1954) analysis of mutation at self-incompatibility loci in higher plants, and to the admirable work of Giles (1956) in Neurospora.

In Drosophila where, thanks to the extensive work of E. B. Lewis (review: 1955), Green and Green (1955, 1956), Mackendrick (1953), Carlson (1957), and others, the recombination analysis is more advanced than in any other higher organism, unfortunately a mutation analysis correlated with the former would find almost insuperable quantitative limitations.

We are therefore left with no alternative but to venture generalisations on the basis of work on three microorganisms: Neurospora, phage, and Salmonella. In the Neurospora work of Westergaard and his group (e.g., Jensen et al., 1951; Westergaard, 1957) and Giles (1951, 1956), the mutation analysis is remarkably advanced, but the recombination analysis not so. It is already clear (Giles, 1951, 1956), however, that in respect of spontaneous or induced back-mutation individual sites of the in (ositol-less) cistron differ markedly from one another.

In phage and Salmonella both the recombination and the mutation analysis have reached an outstanding degree of refinement. A refinement like this is attainable, at a slower but still workable pace, with other microorganisms which have a chromosomal apparatus (Aspergillus and Neurospora, and perhaps yeasts) like that of higher organisms. For higher organisms the outlook is not promising, until genetic analysis based on mitotic recombination in tissue cultures (Chapter VI) becomes a practical proposition.

In Salmonella, Z. Hartmann (1956) found that among eight galactose-nonfermenting mutants (representing probably eight different sites of one cistron) the spontaneous mutation rate towards fermentation was about twenty times greater for the mutant with the highest than for that with the lowest non-zero mutability. Among fourteen serineless mutants in the same cistron this ratio was about 300.

In the analysis of a portion of the $rIIA$ cistron in phage T4, Benzer (1957) found that out of 145 mutants distributed among eleven sites, 123 occurred at one site, and fell into three classes clearly distinguishable by their reverse-mutation rates.

Finally, Streisinger and Franklin (1956) found in phage T4 that the reverse-mutation rates from individual h^+ mutants, representing individual sites of one cistron, to the wild-type h varied over several orders of magnitude.

There is another point worth mentioning. It looks as if the difference in mutation rates between different sites is a property of groups of neighbouring sites rather than of individual sites. In other words, within a cistron there are small segments, extending over several sites, with higher mutability. When we isolate mutants we are more likely, of course, to collect them from these "hot spots" (Benzer, 1957): hence the anomalous distribution of recombination fractions along a cistron noted in Table 2 and earlier: hence also the *non sequitur* in Green and Green's (1956) argument that it is likely that there are only about three sites in the lozenge cistron in Drosophila because the twenty mutants tested seem to be grouped in or around three positions.

Finally, if there is any reality in the picture of the gene as an array of mutational sites, and if there is any reality in the picture of this array as a stretch of DNA duplex in which each site is a nucleotide pair, there is open to experimental probe a very attractive possibility. That is, the use of antagonists specific for each of the four kinds of nucleotide may produce specific mutations at the corresponding sites.

It is conceivable that an approach of this kind may permit the identification, in a well mapped cistron, of the sites sensitive to adenine and thymine antagonists and of those sensitive to guanine and cytosine antagonists. If this were possible, the sequence of nucleotides in DNA—which chemical analysis is at present not able to determine—could be established by genetical means.

Work on this approach is in progress with phage and should be attempted with other microorganisms such as Salmonella and Aspergillus. It would probably require a mapping of over 100 sites in a cistron in order to begin to tell something. However, even if with much less labour it only showed that some sites are sensitive to both adenine and thymine analogs, and others to both guanine and cytosine analogs, the implications would be far reaching enough.

CONTINUITY AND DISCONTINUITY OF THE GENETIC MATERIAL

We have seen that mutational sites within a cistron can recombine as, of course, sites of different cistrons can. This makes the old assumptions about the triple nature of the gene unnecessary, and suggests the adoption of three different units, defined empirically.

However, the question of whether or not there is discontinuity, either or both structural and functional, between one cistron and a neighbouring one, still stands. The problem can now be restated, and it is that of whether genes, though based on more than one ultimate unit of recombination, are still discrete, in the sense that the bonds which recombination resolves between them are of a different nature than those which recombination resolves within them.

We have seen furthermore that the functional criterion of non-complementarity between recessives of one cistron and complementarity between those of different cistrons is not an absolute one. However, it is still conceivable that a cistron usually has definite boundaries, i.e., includes a definite group of mutational sites, none of which are shared with its neighbours, and what seems to be the case in the much quoted example of *ad9* is the exception.

If two contiguous cistrons usually did *not* share mutational sites, recombination between two contiguous sites could still be one and the same process irrespective of whether the sites belong to one or to two cistrons, but the two cistrons would be functionally quite distinct. This is, of course, a possibility which has to be set aside until we have some idea of how to tackle the problem of the primary function of a gene, otherwise than with the empirical *cis-trans* test. In this context, however, it is worth mentioning a possibility.

Brenner (1957), Crick *et al* (1957), and Crick (1958) have developed models of how a structural code made of a continuous array of elements could still determine along its length functional singularities which do not overlap. Their suggestion is in relation to the problem of how a continuous sequence of nucleotides may carry the information for a continuous sequence of aminoacid residues without any overlapping between contiguous nucleotide sequences. I wonder whether the problem of discontinuity at the next higher level—between cistrons—could be approached in a similar way. The problem is that of finding a model which permits discontinuity of activity without structural discontinuity along the sequence of sites at the transition from one cistron to the next. The model would have to suggest how, e.g., a protein or RNA being synthesised along the sequence "knows" where its place starts and where it finishes. My guess is that a good look at the configurations in the lampbrush chromosomes of amphibia, both at the visible and at the electron-microscope level, may give some clue (see Callan, 1955; Gall, 1956).

In any case, the interposition between cistrons of short sequences of elements with no functional significance ("nonsense" sequences, as Crick *et al.* would say) could be sufficient to establish functional discreteness between cistrons without any need for actual discontinuities in the kinds of elements which make up the linear structure of the genetic material.

The absence of any evidence, for or against, leaves it an open question for the time being whether or not there are two different types of bonds along one chromosome, i.e., bonds between cistrons and bonds between sites within one cistron. Mazia (1954) showed that the chelating agent EDTA, presumably by removal of calcium or magnesium, breaks in vitro the chromosome thread into sections of about 4,000 Å., and Steffensen (1955) showed a parallel effect of calcium deficiency in vivo. This might suggest a differentiation, and so does Ambrose and Gopal-Ayengar's (1953) suggestion of longitudinal hydrogen bonds. So far, the work to test these suggestions has been limited and inconclusive (e.g., Levine, 1955; Eversole and Tatum, 1956), and the original interpretation of Mazia is probably based on a fallacy (Kaufmann and McDonald, 1956).

Clearly, if the bonds between cistrons were, e.g., cations and the cistrons were nucleoproteins (or any other distinction which may be more plausible), we may be able sooner or later to find differential effects of specific treatments on recombination between cistrons as compared with that between sites of one cistron. The experimental prerequisites for tests of this kind are already available in Aspergillus.

The problem, however, is in two parts: first, whether or not there are, along the chromosomes, longitudinal bonds of two kinds; and, second, in the affirmative, whether or not the two kinds are inter- and intra-genic respectively.

THE SPECIFICITY OF MUTATIONAL SITES

Keeping in mind that there is a whole range of recombinational and functional relationships between chromosome segments, we can still try to make a picture of a few, arbitrarily chosen, types. The extremes of the range are: 1) pairs of recessive mutants with related or indistinguishable phenotypic effects, located extremely close together (up to the point of no separation by recombination), non-complementary, and showing the *cis-trans* effect; 2) pairs of recessive mutants determining unrelated phenotypic effects, located far apart, and complementary, i.e., the heterozygotes $\dfrac{m_1 +}{+ m_2}$ and $\dfrac{m_1\ m_2}{+\ +}$ are both phenotypically $++$.

One question is how far we can go in differentiating between the effects of mutation at different sites in cases like 1). Here the literature is rich in examples. Among organisms with standard chromosomes we can choose from Mackendrick's (1953 and unpublished) work on w, Lewis's (1955) on bx, Green's (1956) on lz, v, and f, Carlson's (1957) on dp, or Hexter's (1956) on sn, all in Drosophila; from Roper's (1950) work on bi and $paba$, Pritchard's (1955) on $ad8$, Forbes (1955) on pro, Calef's (1957), and Martin-Smith's (1958) on $ad9$ in Aspergillus; and from Giles's (1951) work in Neurospora.

These are all examples, with the possible exception of $ad9$, of mutants belonging to one cistron and bearing to each other relationships of identical, similar, or related phenotypic effects and non-complementarity.

To these examples it would be most instructive to be able to add those from the remarkable work of Stern and his school on the phenogenetics of ci in Drosophila. This would provide, among others, unique information on wild-type "isoalleles" (Stern and Schaeffer, 1943), i.e., alleles at the opposite extreme of the scale from those considered here. However, the work on ci has not been concerned with the differentiation of alleles by means of recombination. Hence there is no way of relating the phenogenetics of the ci alleles to the differentiation of sites in that cistron.

Among organisms with an unknown structural basis for linked inheritance, the outstanding example is that of the $rIIA$ cistron in phage. Others are available in *Escherichia coli* and, thanks to the use of "abortive transduction" by Ozeki (1956), some are on the way from Salmonella.

A good example is that of the $ad8$ cistron in Aspergillus studied by Pritchard (1955, 1958). Here seven recessive alleles of independent origin fall into at least six mutational sites. Five alleles are phenotypically indistinguishable from each other but are distinguishable from two other alleles quantitatively, i.e., on the basis of the extent to which they can grow even without adenine. By tests of recombination, including the recovery of the two complementary products of crossing over from one nucleus (Chapter IV), the alleles can be located in a linear order. There are two known recessive suppressors—one on the same chromosome but over 100 units away, and another one on a different chromosome—which act specifically on one of the alleles but not on the others. The order of the mutants is: $ad20 — ad16 — ad12 — ad11 — ad8 — ad10$ plus one not yet located but certainly left of $ad11$.

The two specific suppressors restore the phenotype of $ad20$ strains to wild type but have no effect on the other alleles. This is similar to what Green (1955) found for the v and f cistrons of Drosophila, and Yanofski and Bonner (1955) for td in Neurospora. On the other hand, Forbes (1955) has found several nonspecific, semidominant suppressors of proline-requirement in Aspergillus which act not only on all the alleles of one cistron, but also on those of a nearby complementary cistron with similar phenotypic effect. Examples of nonspecific suppressors are known in the Drosophila literature.

A significant situation in the case of the suppressor of v is that

while, as Green (1955) has shown, it is site-specific for mutants of the
v cistron, it also "suppresses" mutants of completely different cistrons. This means that the suppressor acts by modifying some cell
condition to which the changed products of several genes are all
sensitive, though only those changed in particular ways. In terms
of enzymes, as gene products, any alteration, e.g., of pH, would
affect all the cistrons within which some mutants produce a correspondingly pH-sensitive enzyme. In terms of sites, the alterations of
this kind need not be based on many, but only on one or a few,
hence the specificity.

Even more interesting examples of specificity are the only two so
far found of partial suppressors within one cistron. They are both in
Drosophila. One (Lewis, 1951) is a partial suppressor within the
Stubble cistron. The other (Mackendrick, 1953) is a partial suppressor within the *white* cistron. In both these examples, the intra-
cistron partial suppressors are alleles which in certain multiple
combinations with other recessive alleles of the cistron determine a
less mutant phenotype than singly.

If the effect of these suppressors within a cistron were complete—
in the sense of restoring the normal phenotype—we would have
here examples of how the wild type can be restored by further mutation within a cistron, though at a site different from the one of the
"forward" mutation. A situation of this kind is likely to occur in
some of the cases of apparent "true" backmutation. It is interesting to speculate on how many combinations of forward and suppressor mutations (besides true backmutations) may be possible in
a cistron. However, in this connection we must remember that
Streisinger and Franklin's (1956) analysis of backmutation in the
h cistron in phage found evidence of backmutation only at the site
of the "forward" mutation.

The case studied by Mackendrick (1953 and unpublished) is as
follows. The "spontaneous" Drosophila mutant w^{aE} (apricot eye
colour, Edinburgh strain) turns out to be mutant at two sites.
Crossing over in a heteroallelic cross $w^{aE}/+$ separates out an extreme mutant phenotype (white).

Mackendrick has also confirmed what was clear from the work of Lewis (1955) and Green and Green (1956), i.e., that phenotypically distinguishable alleles can be obtained by combining or separating by means of crossing over two or more mutant sites, and that the same phenotype may be determined by different combinations of mutant sites. In other words, there can be a variety of interactions between alleles depending on which and how many are combined in one chromosome pair in the two or more possible arrangements between the two homologous chromosomes. Furthermore, "new" alleles arise regularly by crossing over. The results obtained so far are given in Table 7, in which alleles indicated in the same column are those for which evidence of recombination is not available, either because it was not obtained in a certain sample or because it has not been tested.

TABLE 7

ANALYSIS OF THE w CISTRON IN *DROSOPHILA MELANOGASTER*

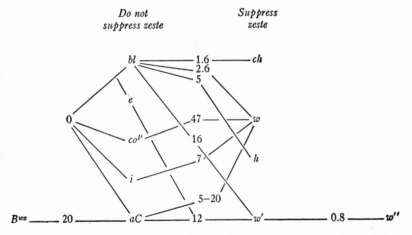

The alleles tested are indicated omitting, in all but three cases, the basic symbol w.

Phenotypes: w = white; w' = white; w^{aC} = apricot (CalTec); w^{bl} = blood; w^{ch} = cherry; w^e = eosin; w^{cP} = coral (Paris); w^i = ivory; w^h = honey; w'' = unknown; w^{Bwx} = brownex; $w^{bl}w'$ = white; $w^{bl}w'w''$ = white; $w'w''$ = apricot (w^{aE}).

The lines connecting the symbols of two alleles indicate which tests of recombination (5,000 to 50,000 progeny) have been done. The figure along each line indicates the recombination fractions $\times 10^{-5}$ obtained.

Source: Mackendrick and Pontecorvo, 1952; Lewis, 1952; Mackendrick, 1953 and unpublished; and Judd, 1957.

The alleles, which may or may not be distinguishable by their effects on colour, can also be classified on the basis of 1) Muller's balance compensation, i.e., identity or otherwise of colour in the hemizygous male as compared with the homozygous female, and 2) dominant suppression of the sex-limited closely linked recessive *zeste* (Gans, 1953).

Mackendrick examined the relations between these two features and the position of each mutant in the cistron. The results are consistent with a difference in position between mutants which suppress *zeste* and those which do not suppress it: the former are all on the "right hand" of the cistron. In view of the existence of site-specific mutants subject to suppression, it is not surprising that there might be regional specificity within a suppressor cistron. The example just given shows regional differentiation, rather than site-differentiation, in the w cistron: it is possible, to say the least, that further work with the three mutants w, w' and w'', which all suppress *zeste*, would show them to represent mutations at three sites. On the other hand, Mackendrick did not find consistent relations between position and presence or absence of balance compensation.

As pointed out by Goldschmidt (1955), the idea of a one-to-one relation between each mutational site of a cistron and a specific effect in the mutants is nothing but the classic particulate gene which would now re-enter through the back door and become equated to a mutational site with the embarrassing addition of the "position effect." This idea is likely to lead to the same fallacies, in spite of some initial apparent successes, as did all previous attempts to find a one-to-one correlation between locus and character. The point is that one cannot use a breeding operation (which defines, by means of recombination, a structural element) and expect it to give one-to-one relations with a physiological analysis (which defines a functional element).

Clearly this idea implies that as we analyse a chromosome segment, we shall resolve it not only into further and further structural elements by crossing over, but also into an increasing mosaic in which each part has a distinguishable specific function which is not merely the consequence of its spatial relations to the others.

As Pritchard (1955) has stressed, the very fact that mutational sites appear to be many, at least in some cistrons and in some organisms, suggests that this approach is unlikely to be fruitful. I should not like to go so far as to say that it is conceptually quite meaningless. By analogy, changes in specific groupings of a molecule have specific effects on its properties: these properties are the result of the molecule as a whole and not of the sum of its parts. Yet it is not wholly unprofitable or meaningless to inquire as to the specific effects that changes in specific groupings have on the properties of that whole. This is in fact one of the hobbies of those working on chemical mutagenesis or on chemical carcinogenesis.

In view of the large array of situations already mentioned, it is possible that we shall find a whole range of situations even in respect of the functional individuality of sites or parts of a cistron. We should add the complications introduced by the structural suppression of recombination by means of minute rearrangements. This can produce systems—"super genes" as Darlington and Mather (1949) call them—like that of the t series in the mouse (Dunn, 1956) of self-incompatibility in plants (D. Lewis, 1954), and several other examples.

A most fruitful result of the reappraisal of the theory of the gene that is going on at present is precisely that we begin to be aware of this variety of situations. In respect of the particular subject of this chapter—allelism—it is clear that neither of the ways in which it can be defined (function or position) has general validity. We can use this concept only as an expedient provided that we know its limitations, and provided that we use empirical and unequivocal criteria to define our terms.

CHAPTER III

STRUCTURE AND FUNCTION OF THE GENETIC MATERIAL

THE theory of the gene assumed autonomy in action of each gene which was supposed to be a distinct small portion of a chromosome linked to others by bonds at which, and at which only, crossing over took place. The beads-on-a-string picture, sophisticated as it may have become later, considered essentially two different entities: the genes and the intergenic connexions. The primary action of each gene was supposed to be independent of the proximity of other genes, though not the secondary and further removed effects of this activity. Position effects were considered to be changes brought about in these secondary interactions by changed neighbourhood (e.g., Muller, 1938).

The preceding chapter gave the reasons why this model is no longer tenable. It is good enough, I repeat, for almost all routine genetics. It is not good enough when we begin to ask what the genetic material is, and how it works.

LEVELS OF INTEGRATION

What are the relations between localised activity of the chromosomal material and its architecture, and in particular its linear differentiation? This is another question which has been in the foreground since the earliest times of genetics.

In the picture which is taking shape at present, the chromosomes cannot be subdivided into units which are at the same time the ultimate units of activity and of crossing over. Rather, the chromosomes are seen as made up of a linear series of elements, or sites, which—in groups of variable numbers—are integrated in activity: mutation is any change in quantity or quality or arrangement of the sites which is still compatible with replication.

The chromosomal section including all the sites integrated for one activity we now call a gene or a cistron. However, we cannot yet say whether or not there is sharp discontinuity between one cistron and another, and if there is, whether it is a general feature of all cistrons: the example of *ad9* given in Chapter II suggests that it is not. It is quite possible that some cistrons overlap or are intermingled with others, and that one and the same site may "belong" to one cistron in respect of one activity and to another one in respect of a different activity. In relation to changes during development, Mather (1948) has suggested something of this sort.

Besides these primary systems of integration there are series of secondary ones. Goldschmidt (1955) has pictured this situation by speaking of a hierarchy of fields. Changing his designations to use the terms which I have used throughout:

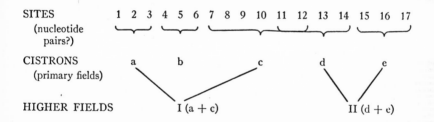

The present picture is much less clear cut than the older one but it lends itself to probing in various directions. In particular, one can imagine experiments capable of disproving the linear continuity of the genetic material, should there actually be discontinuity. Winge (1955) and Taylor (1957) suggested independently that a chromosome may have side branches and that recombination between branches need not involve crossing over along the central axis. This matter will come up again in Chapter IV.

Furthermore, one can imagine experiments for detecting the number of, and the part played by, the individual sites at different hierarchical levels; and experiments on "time of action" (Haldane, 1932) of the genetic material in terms of the epigenetic (Waddington, 1956) emergence of new fields during development or as a consequence of experimentally changed external conditions (Beale, 1954).

It should be clear that in the majority of cases in genetic analysis we are usually considering phenotypic effects which are epigenetically far removed from the primary action of cistrons. And it may well be that a search for this primary action has little meaning, in the same way as it would have little meaning to ask what is the primary action of a particular group of aminoacid residues in the activity of an enzyme molecule.

Let us consider a segment of the genetic material of a chromosome as composed of a structure, the function of which is to supply the information for the synthesis of a protein. The ultimate function of this material would be specified perhaps by the sequence of groups of, say, three nucleotide pairs, each determining which aminoacid is going to join with another to form a dipeptide. But why may it not be the sequence that determines a pentapeptide? Why may it not be the sequence that determines the whole polypeptide chain of the protein?

We have seen that cistrons, i.e., fields of allelism or genes, could be based on a chromosome length of the order of, say, 1,000 nucleotide pairs. The fact that recessive mutants within some such structure do behave as alleles reveals some sort of unitary activity. Unfortunately we have also seen that this unitary activity cannot constitute an absolute criterion. We already know of examples which suggest a gradual passage from one activity, expressed as allelism (i.e., non-complementarity and *cis-trans* effect) of two closely linked mutants, to two activities expressed as non-allelism or complementarity.

GROUPING OF PHENOTYPICALLY RELATED GENES

This said, we can look in the classical way at the relations between arrangement and localised activity of the genetic material. We can then see whether they can be reappraised, perhaps more fruitfully, in the light of more recent views.

We shall start from using, and adding to, the data collected by Grüneberg (1937) and Komai (1950) on close linkage between com‚ plementary genes affecting the same character, or as we should say now, complementary cistrons. Some of the cases in Komai's list can now be removed and classified as cases of allelism (e.g., in Drosophila: *Star-asteroid*; *Stubble-stubbloid*; *bithorax-bithoraxoid*; in the silkworm: the *E* series [Tsujita and Sakaguchi, 1953] and others).

The examples listed in Table 8 are of complementary genes, very closely linked and having phenotypic effects which are indistinguishable or obviously related.

TABLE 8

EXAMPLES OF CLOSE (OR COMPLETE?) LINKAGE BETWEEN
COMPLEMENTARY GENES WITH SIMILAR OR RELATED
PHENOTYPIC EFFECTS

Organism, total linkage map identified, No. of genes mapped	Linkage group	Recombination fractions
DROSOPHILA MELANOGASTER (280, over 500 genes)		
chlorotic, yellow, silver, pallid	I	
yellow 2, yellow N	I	
achaete, scute	I	
lozenge, almondex	I	0.001
lozenge *50e*, other alleles of *lz*	I	
raspberry and vermilion	I	0.008
miniature, dusky	I	0.0005
bright, purple, light, lightoid	II	0.02
bloat, puff, bloated	II	0.003
arch, archoid	II	0.002
blistered, balloon	II	0.001
tetraltera, ophthalmopedia	II	0.003
maroon, rosy, karmoisin	III	0.02
glass, glass-like	III	0.009
cardinal, white ocelli	III	0.005
ASPERGILLUS NIDULANS (660, 40 genes)		
pro1, pro3 (prolineless)	I	0.005
ad15, ad17 (adenineless)	I	0.0001
ad1, ad3 (adenineless)	II	0.001
s1, s12 (requiring sulphite)	III	0.03
NEUROSPORA CRASSA (800, 80 genes)		
al1, al2 (albino)	I	0.02

TABLE 8 (Continued.)

Organism, total linkage map identified, No. of genes mapped	Linkage group	Recombination fractions
arg3, B312 (arginineless)	I	0.03
hist2, hist3 (histidineless)	I	0.05
pdx, pdxp (pyridoxineless)	IV	
iv1, iv2 (isoleucine-valineless)	V	0.03
pan2 B3, pan2 B5 (pantothenicless)	VI	0.005
am32, am47 (no glutamic dehydrogenase)	?	0.0000

SALMONELLA (?, 42 genes)
tryA,B,C,D
hiA,B,C,D
cysC, cysD
(details of three other examples referred to by
Demerec, 1956, have not been published)

ESCHERICHIA COLI (2,000, 20 genes)		
ind, Lac1, Lac4 (lactose permease and lactases)	—	0.001
Gal1, Gal2, etc. (UDP-transferase and galactokinase)	—	0.001

PHAGE T4 (800, 10 genes)		
rIIA, rIIB (plaque morphology)	—	0.02

MOUSE (1954, 50 genes)		
T, Fu¹ Ki (brachyury and other tail abnormalities)	IX	0.05
T, t⁰-t⁴⁴, tʷ'tʷ¹⁷ (brachyury)	IX	0

SILKWORM (338, 56 genes)		
p, S (larval patterns)	II	0.02
E, Nc (extra larval legs, reduced semi-lunar patterns)	VI	0.01
ae, be (amylase in body and digestive fluids)	VII	0.01

COTTON
M, G, S (presence and distribution of anthocyanins)   ?

References: Drosophila—Bridges and Brehme, 1944; Slatis and Willermet, 1953; Komai, 1950; Aspergillus—Käfer, 1958; Neurospora—Barratt et al., 1954; Newmeyer, 1957; M. Mitchell, 1955; Chase and Giles, 1957; Fincham and Pateman, 1957; Salmonella—Demerec et al., 1956; Demerec, 1956; Escherichia coli—Lederberg, 1951; Morse et al., 1957; Kalckar et al., 1956; Phage T4—Benzer, 1955; Mouse—Dunn, 1954, 1956; Silkworm—Tanaka, 1953; Cotton—Yu and Chang, 1948.

Do these examples suggest some orderly relation between position and effect of genes? For a number of reasons it is difficult to state the question precisely for a statistical analysis. Attempts to determine whether the distribution of genes along and between chromosomes is not random in respect of their effects are not new, and one of the earliest was by Dunn and Caspari (1945).

Let us take again the example of the adenineless mutants in Aspergillus reported earlier. Out of a total of eight identified loci at which recessive mutants determine adenine requirement, two (*ad1* and *ad3*) are very close together (0.001 recombination). With a total map of more than 660 units assuming the distribution of identified mutants to be random, the probability that one adenineless mutant is located in a segment 0.1 of a unit on either side of another adenineless mutant to which it is complementary, would be $2 \times 0.1/660 = 3.3 \times 10^{-4}$.

Since there are seven groups of adenineless mutants identified as complementary in addition to the one in question, the probability that the one in question be as near as 0.1 units to any one of the other seven is $7 \times 3.3 \times 10^{-4}$. The probability that any one of the eight is near any other one is $28 \times 3.3 \times 10^{-4} = 0.009$. The same argument could apply to two complementary proline requiring mutants (recombination 0.005) out of two groups identified (Forbes, 1955). The probability here would be $2 \times 0.5/660 = 0.001$, and so on.

In other words, the two known examples in Aspergillus of very close linkage between complementary genes with related effects cannot be due to chance if the distribution of genes is random in relation to their phenotypic specificity.

The examples from Neurospora are even more striking than those from Aspergillus: there are seven known pairs of closely linked complementary genes determining different blocks in the same chain of reactions. In one case (*am32* and *am47*) the two mutants seem to determine a loss of the same enzyme activity (Fincham and Pateman, 1957).

Only a much broader test on many different groups of mutants in many organisms could give a precise idea of how big this departure from randomness is. The impression from the data on Table 8 is that there is a rather strong tendency for genes related in effects to be near to one another.

There are cases, of course, in which the grouping of genes with related functions is obviously the consequence of a special situation.

For instance, the distal portions of both long and short arms of the Y chromosome in *Drosophila melanogaster* are required to be intact for sperm motility in the male. Neuhaus (1939) has identified at least fourteen positions at which mutation determines a recessive failure of motility, all complementary to each other. Clearly one of the solutions, in fact the one mechanically simplest for a chromosomal determination of a male sex-limited property, is that of a specialized Y chromosome. In this case the linkage need not be relevant to the functional dependence on proximity of the genes involved but only to the mechanisms of hereditary transmission.

The question of how, in the evolution of the chromosomes, arrangements like this or of any other kind may have arisen is an interesting one but not the one which we are considering here. We are enquiring into the relevance of the position of genes to their function rather than into the historical processes which have brought about any non-random arrangements of genes. Not keeping these two points distinct has led to confusion such as the idea that "pseudo-allelism," as it was called, or allelism as it should be called, is the consequence of duplication of genes (Lewis, 1951). Though allelism may have arisen after duplication, at least in some cases, this does not help at all in understanding why two recessives should show functionally allelic relations.

We have to keep in mind again that when we are attempting to group genes according to their effects, we do so at a phenotypic level which is likely to be very far removed from what happens at the chromosome surface, and there might be more order at this ultimate level than meets the eye. All told, however, we can say that though in several organisms there are many more cases of proximity of genes with related effects than would be expected by chance, the proportion of these cases is relatively small and does not suggest any generalisation.

That no simple generalisation is possible arises also from consideration of "suppressors." These are mutants restoring the normal phenotype when in combination with other complementary mutants which by themselves determine a departure from normal. The distribution of suppressors does not seem to bear any spatial relation to the chromosome sections which they "suppress." This is probably a more objective test of functional relationships than one merely based on similarity of phenotypes.

The case of Salmonella, studied by Demerec and his collaborators (1956), requires consideration by itself even conceding that tests of linkage based on transduction may be open to reappraisal. The examples published are quite striking. Mutants determining nutritional requirements for tryptophane, histidine, cystine, adenine, and leucine (presumably because blocked in the syntheses of these metabolites) are grouped in a very specific way. Four cistrons, each with many mutational sites already identified, determine tryptophane requirement with blocks at four successive steps in its synthesis: these four cistrons are adjacent to each other and in a linkage order corresponding to that of the temporal sequence of steps in the synthesis. The same is true for four cistrons in which mutants determine requirement for histidine. For nine steps in the synthesis of purines, of which blocked mutants are available, there are nine cistrons, two and three of which, respectively, are closely linked. For four steps in the synthesis of leucine, the two corresponding cistrons are closely linked. Finally, two out of five cistrons, mutants in which control five steps in the synthesis of cystine, are closely linked. For seven other gene-controlled sequences of essential metabolites the corresponding cistrons are not linked. In a total of twenty-four cistrons involved in five sequences of reactions, seventeen are grouped; and eighteen other cistrons involved in seven other sequences are not grouped.

This is undoubtedly an impressive example of non-random arrangement of cistrons. It is precisely of the type postulated (Pontecorvo, 1950) in the now untenable picture of an assembly line of genes controlling successive steps in biosynthesis of an effectively limited metabolite. It is not clear, however, what part in it is played by the method of assessing linkage, i.e., transduction. We do not know if the transduction method is comparable in resolution to that based on meiotic or mitotic recombination. For instance, in

Aspergillus (Table 5), if the map were full of blanks and recombination very low, we would consider as closely linked four out of the eight known cistrons controlling steps in purine synthesis, and three out of four cistrons controlling steps in the reduction of sulphate.

However, as has been suggested by various authors, it is also conceivable that these examples of linkage in Salmonella imply that functions carried out by special cytoplasmic organelles in higher organisms are carried out in assembly-line fashion by chromosome-like structures in this bacterium. In other words, the delegation of routine metabolism by the nucleus to the cytoplasm has not progressed much in this group.

The close proximity of genes controlling related or sequential biochemical steps, which is the exception rather than the rule in other organisms, is less of an exception in Salmonella.

The preceding analysis of close linkages between complementary genes related in function raises one point which does not seem to have been appreciated. In genetic mapping one encounters close linkages more often than would be expected on the basis of a uniform distribution of genes (Pontecorvo, 1956). This affects pairs of genes obviously related in action (like those controlling successive steps of a biosynthesis) as well as pairs of genes not so obviously related.

For instance, the 1957 map of the eight chromosomes of *Aspergillus nidulans* (Table 5) is at least 660 units in length. There are about forty identified loci and among them as many as five pairs of loci are 0.5 or less units apart: *ad1* and *ad3*; *pro1* and *pro4*; *ad9* and *pabal*; *y* and *ad8*; *pyro* and *orn4*. The same situation is apparent from a look at the linkage maps of organisms as diverse as the mouse and Neurospora. What underlies it?

Two suggestions come to mind: 1) a non-random distribution of crossing over, with a number of small chromosome segments in which its incidence is well below average; 2) a non-random distribution of mutation rates, with localised small regions of high mutability.

The former is quite possible but so far there is no evidence either way. In the latter case the excess of close linkages would be a reflexion of the fact that the markers which one uses are more likely to come from regions of high mutability. We have already seen that this is a likely explanation for the parallel excess of very close linkages found between sites at the intra-cistron level.

On the whole the excess of close linkages, be it between genes related in action or otherwise, could well be mainly another aspect of the fact that linkage maps give a highly selected picture of the genome.

FIELDS OF HIGHER ORDER

We have seen that by and large there is no more than a tendency towards an assembly-line arrangement of cistrons affecting the steps in a sequential process (Pontecorvo, 1950). We have now to ask how widespread and important may be spatial integrations involving chromosome regions further apart: the "higher order" fields of Goldschmidt (1955). The question here is not concerned with epigenetics but with structural integration at the chromosomal level. A theoretical discussion by Sewall Wright (1953) is very pertinent in this context.

The information available is very meagre. The best known examples of higher fields, apart from sex determination, are the homoeotic mutants in Drosophila and the brachyury system in the mouse. It is possible, in addition, that what McClintock (1956) calls "controlling elements" in maize are precisely examples of "higher fields."

The mutants in Drosophila are polycomb 45, proboscipedia 48, tetraptera 51, aristapedia 49, bithorax 59, all located in the third chromosome in a section of about fifteen units, and all affecting segmentation and segmental homologies. Others in Drosophila are tetraltera and ophthalmopedia in the second chromosome, three units apart.

The case of *Ki, Fu,* and *T* in the mouse, studied by Dunn and his collaborators, in particular Gluecksohn-Waelsch, is another perhaps more striking example: it has been reviewed recently (Dunn, 1954, 1956) and for details the reader is referred to these reviews. There is here a system of genes (*Ki*, *Fu* and *T*) extending over about five units, the mutants of which block differentiation in the posterior axial system at various stages. Two of the three have been separated by crossing over.

Furthermore, the *T* series by itself is an example of a very complex situation at a level which it is difficult to classify as inter- or intra-genic. A large number of mutants, mainly recessive, which are complementary to each other is known: the test of complementarity tells that so far about thirty are different. The test of recombination, on a substantial basis, has not resolved them. This is probably only a foretaste of what we are likely to find when we pass from the analysis of "simple" genes, like those providing the information for such a minor matter as the mere synthesis of an enzyme (probably represented by most of the examples on Table 6), to the genetic organisation necessary to carry the information for morphogenetic processes.

It is a sobering reflexion that while we begin to have some speculative ideas of the amount and the kind of information required for protein synthesis (e.g., Crick, 1958), we have no idea of what may be required even for the simplest of morphogenetic processes.

Granted that the present knowledge of "higher fields" is meagre, it could be argued that the routine techniques with which we isolate mutants are of a type which tend not to pick up cases of this kind. Perhaps an analysis of the genetic fine structure and developmental effects of lethals could reveal some of these integrations at a distance. The phenogenetic basis begins to be available (Hadorn, 1955).

In routine isolation of mutants we have no means of identification for a system of the following kind, which is one of those expected in a hypothetical case of integration at a distance: 1) a recessive mutant *a* in a region of a chromosome produces a certain effect; 2) a recessive mutant *b* differently located, or even in a different chromosome, produces an indistinguishable effect; 3) the two act as functionally allelic, in the sense of being noncomplementary: the double-heterozygote has a mutant phenotype.

Apart from the cases described by McClintock in maize, the only clear example of this kind is the one of transition from allelism to complementarity in the *ad9* system in Aspergillus mentioned earlier. In this example, however, the recessives *a* and *b* (corresponding, e.g., to *ad13* and *ad32*) are extremely closely linked and it is only the knowledge that other sites, e.g., *ad17* and *ad15*, with different functional relations are interposed between them that makes the case unusual.

If only two mutants, e.g., *ad17* and *ad15*, had been investigated, this would be only an additional example, with nothing unusual, to be added to the list (Table 8) of closely linked complementary recessives, indistinguishable phenotypically.

Perhaps there is a maximum critical distance for allelic relations to operate, and this could be just less than the mean distance between two points within the resting nucleus, each on one of two homologous chromosomes. There is nothing on which to base speculations about this distance, because we do not even know what structure and arrangement the chromosomes have in the interphase nucleus.

In this respect there is a point worth considering, and I am indebted to Dr. Pollister for having called my attention to it. We do not know what sort of arrangement the chromosomes have in the interphase nucleus, in which presumably the genetic material is in active state. It is possible that the chromosomes are not distributed in a disorderly way within the nucleus and are not moving about at random. They may have orderly foldings and regions of "ectopic" juxtaposition (Slizynski, 1945), both along one chromosome and between non-homologous ones. There may even be some sort of repetitive juxtaposition like that along successive gyres of a helix. All this might result in proximity between segments quite apart along the chromonema, or even belonging to different chromosomes. In diploids or higher polyploids (but not in haploids) juxtapositions involving different segments of one chromosome pair could take place both between two segments of one member of the pair or between two segments, one on each of the homologous chromosomes.

Techniques for detecting structural-functional relationships of this sort have not yet been worked out, but they should not be impossible to design. The crucial question would be whether between two segments far apart on one chromosome, or on different chromosomes, there may be relationships similar to those of allelism. It will be worth while to start thinking of ways to investigate these problems. The "higher fields" could have a structural basis quite different from the one earlier implied, which is essentially a "supergenic" linear arrangement along one chromosome or part of a chromosome. They could be based on structural integrations between chromosomes, or chromosome parts, which are not revealed by ordinary crossing over analysis.

The outstanding work in maize (review: McClintock, 1956) on the relations between genes, at more or less fixed chromosomal positions, and transposable elements which determine phenotypic changes and transmissible changes, is probably very relevant here. Unfortunately I do not understand the details of this work well enough to put my finger on what may be particularly significant.

In McClintock's words "Controlling elements appear to reflect the presence in the nucleus of highly integrated systems operating to control gene action." In some at least of the systems analysed by McClintock two controlling elements, at widely different (and changeable) positions in the linkage map, interact to determine both the activity and the mutability of one or more particular genes. Apparently the "controlling elements" have no direct phenotypic action, but they affect that of the genes.

· Any one of these systems reveals itself only under very special circumstances; hence they passed unnoticed in maize for a long time. Now that some of these circumstances are understood, new systems are added to the list rapidly. It is conceivable that they are of general occurrence and importance and that they have not yet been discovered in other organisms precisely because one did not know how to look for them. However, it is also conceivable that they are peculiar to maize and, as Dr. Mangelsdorf has explained to me, that they result from the presence in the maize genome of partially homologous duplications derived from the Teosinte genome.

Evidently some of the "higher fields," postulated above, may be systems like those discovered by McClintock in maize. Furthermore, there is a connexion worth considering. We noted the enormous increase between lower and higher organisms in DNA per unit map. Let us keep in mind that this increase might reflect that of "controlling elements." In other words, the number of genes, and therefore the number of different proteins, need not be several orders of magnitude greater in a mammal than in a bacterium, but the number of "controlling elements" may be. Evolution may have taken place by finer adjustments of reactions rather than by an increase in the number of different molecular species.

If there is anything in this idea, the difference in the amount of DNA per nucleus between, e.g., Urodela and Anura, or between Dipnoi and other fishes, could represent an orthogenetic trend which has led almost to the extinction of the Urodela and the Dipnoi. Too much of "controlling elements" may be at least as embarrassing as too big a size or too long a tooth or too branched an antler.

A discussion on the relations between arrangement and activity of the chromosomal material should, of course, include a consideration of heterochromatin. The trouble here is that the study of heterochromatin is at a prescientific level (e.g., Pontecorvo, 1944). We have no alternative but to ignore it.

CONCLUSIONS

For the time being, we must restrict a working model of the genetic material to the level for which there begins to be some solid ground, i.e., the intra-genic level. Summarising again:

1) The chromosome is subdivided in sections within which recessive mutants are noncomplementary to one another. These sections of allelism—the genes of classical genetics—are called cistrons (Benzer, 1957).

2) A proportion of the mutants of a cistron represent changes at individual positions which can be recombined by crossing over and which are in a linear arrangement; the positions within a cistron separable by crossing over are called (mutational) sites (Pontecorvo,

1952); pairs of recessive mutants at different sites show the *cis-trans* difference in phenotypes when combined in a heterozygote (Lewis, 1945).

3) Within one cistron the occurrence of mutations at the same site, i.e., not recombinable by crossing over, is uncommon.

4) The fact that independently arisen mutants of a cistron often turn out to represent mutations at different sites, and calculations based on the ratio of the total recombination in a cistron to the minimum between two sites of that cistron, suggest that a cistron may be based on hundreds of mutational sites separable by crossing over.

5) The non-complementarity of mutants of a cistron and the *cis-trans* effect, which denotes highly localised functional integration, suggest a unitary activity of the cistron as a whole.

This last conclusion, as stressed in the previous chapter, should not be taken as absolute: there are intermediate situations between non-complementarity and complementarity. On the whole it looks as if the beginning of an insight into the relations between arrangement and activity of the genetic material is more likely to be found at the ultimate level of organisation—that of sections of unitary activity, the genes or cistrons.

Is there any working model of a gene which we can make more definite than that of a linear series of sites integrated into a specifically active section? Schmitt (1956) has proposed one such picture, based on the ideas which derive from the relations between structure and activity in collagen, myosin, and other biologically active fibrous proteins. In these the full chemical and biological activity is the property of supramolecular units, say 20 Å thick and several thousand long, which result from specific adlineation of macromolecules. Schmitt's idea is that the genes are not individual macromolecules or parts of individual macro-molecules but that they result from the specific juxtaposition of two or more macro-molecules.

According to Schmitt a number of specific juxtapositions could arise, as they do in the aggregations which form collagen, as a consequence of a number of specific conditions in the nucleus. In differentiation new juxtapositions would arise as a result of the previous history of the cell and would in their turn determine conditions for further juxtapositions.

This last is, in more precise chemical terms, the same idea which I tried to convey very crudely some years ago, that is that new integrations (i.e., genes) would arise epigenetically as a consequence of minor changes in chromosome spiralisation resulting from previous activities. These changes were supposed to bring about new juxtapositions of sites in the chromosomal material.

Another model, which again emphasizes supra-molecular organisation, is that proposed by Kacser (1956). He suggests for the chromosome the structure of an adsorption complex with an interface generated by the complementary surfaces of the two moieties, which are paracrystalline DNA and parocrystalline protein respectively. The part that matters, from the point of view of genetic activity and specificity, is the adsorption surface. Kacser's model also envisages a plausible mechanism of self-replication for the specific adsorption interface.

I am sure that ideas like Schmitt's and Kacser's will have decisive importance in the future. Up to the present, however, a more heuristic model of a gene which has been devised is that derived on the one hand from the ideas of Watson and Crick (1953) on the structure and replication of DNA, and on the other hand from ideas based on the analysis of the fine structure of minute sections of the genome mainly in phage, Aspergillus, Drosophila, and Salmonella.

Briefly, this model is that of a duplex DNA helix some 1,000 necleotide pairs long carrying in the sequence of nucleotides the information for the synthesis of a protein some 300 aminoacid residues long. With this model, the *cis-trans* effect is the consequence of the fact that between two differently damaged homologous helices there is no possibility of reciprocal repair in the act of transfer of the information; i.e., two differently defective proteins are synthesized. This is more or less the same as saying that the information of each of the homologous structures is transmitted *en bloc*. Thus, no *trans* effect is expected nor is it, in fact, found, when we recognise the activity of the gene by identifying the proteins themselves as, presumably, we do (Table 9) with antigens (e.g., Barish and Fox, 1956), enzymes (e.g., Horowitz and Fling, 1956; Fincham, 1957) and haemoglobins (e.g., Pauling, Itano, Singer and Wells, 1949; Ingram, 1957).

TABLE 9

EXAMPLES OF DIFFERENCES IN PROTEINS, INCLUDING LOSS OR CHANGE OF ENZYMATIC ACTIVITY, DETERMINED BY DIFFERENCES IN SINGLE GENES

Organism and condition	Enzyme or other protein	Nature of the detected change
MAN		
sickle-cell anaemia	haemoglobin	substitution of one valine for one glutamic acid residue
MOUSE		
haemoglobins	haemoglobin	electrophoretic mobility
SHEEP		
AB haemoglobins	haemoglobin	electrophoretic mobility
NEUROSPORA		
1. amination deficiency	glutamic dehydrogenase	a) loss of activity
	glutamic dehydrogenase	b) temperature sensitivity
2. arginineless	arginosuccinase	loss of activity
3. tyrosinase	tyrosinase	a) loss of activity
	tyrosinase	b) thermostability
4. isoleucine-valineless	dihydroxyacid dehydrase	loss of activity
5. nitrateless	nitrate reductase	a) loss of activity
	nitrate reductase	b) pH sensitivity
6. tryptophanless	tryptophan synthetase	a) loss of activity
	tryptophan synthetase	b) thermostability
ESCHERICHIA COLI		
1. gal1	UDP-transferase	loss of activity
2. gal2	galactokinase	loss of activity

References: Man—Ingram, 1957; Mouse—Gluecksohn-Waelsch et al., 1957; Sheep—Evans et al., 1956; Neurospora—1a. Fincham, 1954; 1b. Fincham, 1957; 2. Fincham and Boylen, 1957; 3a and b. Horowitz and Fling, 1956; 4. Myers and Adelberg, 1954; 5a and b. Silver and McElroy, 1954; 6a and b. Yanovsky and Bonner, 1955; Escherichia coli—1. Morse et al., 1957; 2. Kalckar et al., 1956.

Numerous other cases of mutational change or loss of enzyme activity in E. coli and Neurospora (e.g., Davis, 1955) and the differences in lactoglobulins (Aschaffenburg and Drewry, 1955) in cattle, are not included because not yet analysed genetically to a sufficient extent. In the many known cases of genetically determined differences in antigens, the nature of the difference has so far been established only immunologically.

It is also implicit in this model what some of the cases of partial dominance may imply. They would be cases in which one allele carries the information for a protein structure different from the other, and both proteins affect the phenotype rather than only one being active. Outstanding examples of this are sickle-cell anaemia in man (Pauling *et al.*, 1949) and the heterokaryons for normal and thermolabile tyrosinase in Neurospora (Horowitz and Fling, 1956). In cases like this it would be most interesting to compare a *trans* heterozygote for two differently altered proteins with the corresponding *cis* heterozygote. The prediction is (Pontecorvo, 1955) that the *cis* heterozygote should in some cases produce two molecular species, one of which is normal: the properties of the other would not be predictable from those of the two proteins in the *trans* heterozygote.

Whatever the merits of a working model like this, it has already stimulated experiments of great elegance like those of Levinthal (1956) on the mechanism of replication of bacteriophage, and those of Plaut and Mazia (1956) and Taylor (1957) on the mechanism of replication of the chromosomes. Thus it is a fruitful model, even if it is an oversimplification. It may be good enough for the analysis of the simplest systems of genetic information, such as those required for the synthesis of an enzyme or other protein. We should not expect it to work in this simple form when the information required is the vastly larger one presumably demanded by a morphogenetic process.

CHAPTER IV

RECOMBINATION

THE analogy was suggested that recombination is to genetic analysis what diffraction is to microscopic analysis, i.e., the phenomenon on which the analysis is based. The analogy stops there: we are in the appalling position that our knowledge of the mechanisms of recombination is very limited. To be sure, we can define precisely recombination, namely, the formation of a genome in which parts originating from different cell lineages have become reassorted. We have also a substantial understanding of one of the processes of recombination—that which Mendel discovered. At meiosis, or even at mitosis in certain cases (Chapter VI), the members of nonhomologous chromosome pairs segregate more or less independently between the two poles. We know now, however, that this process is only one, and a relatively unimportant one, of those at work. Beyond this, the knowledge of the processes of recombination is largely conjectural: this applies in particular to the most important of them, crossing over.

Among the processes of recombination some, like meiotic or mitotic crossing over, are definitely reciprocal, i.e., one event of recombination between two homologous chromosomes yields two chromosomes between which corresponding parts have been reciprocally exchanged. Others are, or at least appear to be, nonreciprocal: transduction and transformation in bacteria, gene-conversion (Lindegren, 1953; Roman, 1956; M. Mitchell, 1955; Strickland, 1958) in organisms with standard meiosis, recombination in bacteriophage (Levinthal, 1954). In these cases only one recombinant product is recovered from one event of recombination. The two complementary products may be obtained, but only from separate events.

The distinction between reciprocal and nonreciprocal recombination is based at present mainly on the failure to recover the complementary products from one event. In some cases there has been a tendency to extrapolate from this negative kind of evidence back to the nature of the process of recombination itself, e.g., in phage.

Obviously it is a fallacy to infer that a process of recombination does not involve reciprocal exchanges on the basis of failure to recover or recognise the reciprocal products of one event. For instance, in oögenesis only one of the four products of meiosis forms the egg nucleus. Thus, the complementary products of crossing over are not usually recovered in one zygote. Yet, by using, for example, attached-X Drosophila, they can be recovered (e.g., Beadle and Emerson, 1935). They can also be recovered in mitotic crossing over (Roper and Pritchard, 1955), in gametogenesis of polyploids, and they are recovered in the overwhelming majority of cases in tetrad analysis as will be shown later.

Most of the types of recombination in which the reciprocal products of one event are not recovered have come to light in recent years. They raise a number of important questions which will be considered in this chapter. The central one would be whether recombination of linked genes has fundamentally the same basis from phage to man, and the obvious differences are either spurious (cf. Emerson, 1956) or not essential. It is no use, however, to ask a question too big for experiment.

In spite of this, we should be well aware that the attempts to unravel the various processes of recombination may start from two opposite working hypotheses: unity or multiplicity of basic mechanisms. To my mind the hypothesis of unity is more heuristic.

In the discussion at the Oak Ridge Symposium on Genetic Recombination in 1954, Crick (1955) made the justified remarks that there is an appalling lack of knowledge on the structure of the chromosomes as soon as we get above the molecular scale, and that geneticists are unable to state clearly what the problems of chromosome replication and recombination are which have to be explained in terms of physical chemistry.

From the geneticist angle, Sturtevant (1951) had earlier made substantially the same complaints, though applied to the more limited field of the details of meiosis, and concluded: "We shall have to wait for the cytologists to produce a more coherent and satisfactory picture before we can hope for an inclusive description of the process of meiosis, and until that is available it hardly seems possible to plan any experimental analysis of the physical forces at work. Meanwhile there remains much for the geneticist to do in studying the mechanics of segregation and crossing over by his own precise methods."

In the following sections I shall summarise some of the recent results obtained by the geneticist's "precise methods," which throw new light on recombination, and hope that they may help to state more clearly what the problems are. These results come from the following types of investigation: 1) recombination analysis in tetrads from heterozygotes with many intervals marked in one chromosome pair. This has been done extensively only in Neurospora (Perkins, 1956) and *Aspergillus nidulans* (Strickland, 1958*b*). In the case of Neurospora it has been preceded by extensive backcrosses to eliminate structural and other heterogeneity. In the case of Aspergillus, all the strains used stemmed from a single one; 2) recombination analysis of minute chromosome segments. This shows remarkable consistency between organisms as different as phage, *Escherichia coli*, Salmonella, *Schizosaccharomyces pombe*, *Aspergillus nidulans*, and Drosophila; 3) analysis of mitotic crossing over, almost exclusively in *Aspergillus nidulans*.

· Simultaneously with the new information from the genetic angle there has been a fruitful crop of new approaches from the physical and biochemical angle: 1) the study of the relations between synthesis of DNA and inferred time of chromosome duplication in both the mitotic and the meiotic cycles (e.g., Pollister *et al.*, 1951; Howard and Pelc, 1951; Taylor, 1953). The main conclusions (Taylor, 1957) are that DNA and chromosome duplication coincide, that they occur before prophase both in mitosis and meiosis, and that therefore any theory of crossing over which requires chromosome duplication after zygotene pairing (e.g., Belling, 1931; Darlington, 1937) is untenable; 2) the study of the mode of duplication of genetic structures (phage, chromosomes) based on the distribution of (labelled) atoms of an existing structure between "daughter" structures (Watson and Maalöe, 1953; Levinthal, 1956; Stent, 1958; Plaut and Mazia, 1956; Taylor, 1957).

TETRAD ANALYSIS: THE RECOVERY OF RECIPROCAL RECOMBINANTS

In organisms with cytologically observable chromosomes and with standard meiosis, the complementary products of exchanges occurring at meiosis in one nucleus can be recovered by the use of special techniques—attached-X Drosophila, polysomics, and tetrad analysis. The complementary products of mitotic crossing over can also be recovered together (Table 14) in certain cases.

Tetrad analysis is possible in a number of organisms in which the four products of a single nucleus which has gone through meiosis are kept in a group. There is a vast literature to which I shall not refer (see review by Perkins, 1955). A point to be stressed is the extraordinary precision of the results of meiosis which tetrad analysis brings out.

First, an overwhelming proportion of tetrads shows normal 1:1 segregations, i.e., for each heterozygous locus, two nuclei of each tetrad carry one allele and two carry the other. Second, an overwhelming proportion of tetrads in which crossing over between linked loci has occurred show for each exchange two, out of the four, products of meiosis carrying the correspondingly exchanged segments. Third, in tetrads in which more than one exchange has occurred, the multiple exchanges may involve only two or three or all four products, but only two of them are involved at each point of exchange.

All these features had long been inferred from genetic analysis based on random samples of products of meiosis as well as from cytology. Tetrad analysis brings home vividly how correct these inferences were.

I shall give here a few examples from Strickland's (1958*b*) extensive tetrad analysis in *Aspergillus nidulans* to support this contention. In this species the four nuclei arising from meiosis divide mitotically again and the eight are then included each in a haploid ascospores; the eight ascospores (four genotypes) are kept together in an ascus. Isolation of the ascospores from the asci requires the micromanipulator. In one of Strickland's crosses 577 asci were classifiable out of a sample of 611 from zygotes marked as follows in chromosome I:

	pro3	+	+	+	+	*bi1*
	+	*pro1*	*ad15*	*paba1*	*y*	+
Standard map units		0.5	8	0.3	16	6
Interval		I	II	III	IV	V

Out of these 577 asci only two had unexpected constitutions. In both, seven out of the eight spores germinated and the genotypes were as follows:

Ascus 16

pro3	+	+	+	+	*bi1*	(3 spores)
pro3	+	+	+	*y*	+	(2 spores)
+	*pro1*	*ad15*	*paba1*	+	*bi1*	(1 spore)
+	*pro1*	*ad15*	*paba1*	*y*	+	(1 spore)

Ascus 17

pro3	+	+	+	+	*bi1*	(2 spores)
pro3	+	+	+	+	+	(1 spore)
+	*pro1*	*ad15*	*paba1*	*y*	*bi1*	(1 spore)
+	*pro1*	*ad15*	*paba1*	*y*	+	(3 spores)

As Strickland points out, these two unexpected asci could have been normal asci, broken and with one spore lost. One of the three spores *pro3 bi1* of Ascus 16, and one of the three spores *pro1 ad15 paba1 y* of Ascus 17, could have been intruders, i.e., picked up accidentally from the mass of free ascospores during dissection. Spores of each of these fully parental genotypes constitute about 35 percent of the total in this cross. Asci in which only six out of the eight spores germinated were $136/611 = 25$ percent in this cross. The product of these two probabilities is about 9 percent for each of the two cases. We cannot gauge how often, in dissection, a broken ascus loses one spore and picks up another one from the spore suspension surrounding it. But it looks as if between the maximum probability of 9 percent possible for the occurrence of each of these two asci and the actual $1/577 = 0.17$ percent with which they occurred in this experiment, there is a sufficient margin for a technical mistake.

Some such argument has permitted Strickland to ignore 11 out of 17 unusual asci found in a total of 1,642 fully classified asci from nine crosses (of which that given above is one) involving a total of fifteen loci. The remaining 6 unusual asci constitute less than 0.4 percent of the total. Emerson (1956) has stressed that before taking unusual tetrads seriously, one should eliminate all the spurious cases or those which can be explained as special cases on classic theory. We shall see that the 6 unusual asci remaining after selection are interesting enough without the chaff of the other 11 probably spurious ones. One good case of abnormal tetrad is more fruitful than dozens disputable.

We shall ignore asci 16 and 17 and look instead at the other 575 of the above cross. They were classifiable as follows in respect of recombination between the markers:

Tetrads with no exchange		340
Tetrads with one exchange		
in I	3	
in II	49	
in III	2	
in IV	105	
in V	32	191
Tetrads with two exchanges		
in II and IV	18	
in II and V	4	
in IV and V	8	
both in II	3	
both in IV	5	38

Tetrads with three exchanges		
in II, IV and V	3	
two in II and one in V	1	
one in II and two in V	1	5

Tetrads with four exchanges		
one in II, one in V and		
two in IV	1	1
		575

In every one of the 235 asci in which exchanges were recognisable, the two reciprocal products of each exchange were represented: this means that the numerical data and the genotype of the zygotes given earlier, contain all the information required for reconstructing every one of the 575 tetrads. Particularly interesting are the three exchanges in interval I and the two in interval III, both less than one unit long. In all five of these tetrads the reciprocal products were present.

This remarkable precision of crossing over even down to minute regions is something with which some of those engaged in the analysis of the newer types of recombination (transduction, phage recombination, etc.) are not quite familiar. Only three years ago, in my laboratory, one of the most brilliant investigators of phage recombination, on a fleeting visit, spent a whole afternoon examining with fascination the protocols of an experiment on tetrad analysis.

INTERFERENCE

Classification of tetrads as to no-exchange, one, two, etc., exchanges, is an arbitrary affair, as Perkins (1953, 1955), Whitehouse (1956), and others have amply discussed. Certain types of exchange cannot be classified with certainty, e.g., within one interval three-strand doubles are indistinguishable from single, and two-strand doubles indistinguishable from no-exchange. To make allowance for this one has to make groundless assumptions on the mechanism of crossing over, which, incidentally, beg the question.

This applies in particular to the analysis of chromatid and chromosome interference. "Interference" was the term introduced by

Muller (1916) for the fact that in Drosophila and maize, recombinant gametes with two exchanges, one in each of two small adjacent
intervals, are produced less frequently than expected from the product of the incidences of recombination in each of the two intervals.
This was interpreted as a structural hindrance of one exchange on
another one nearby.

Haldane (1931) applied the same idea to the coincidence of
chiasmata as measurable by cytological observation ("cytological"
interference) and pointed out that there are two types of interference, since referred to as "chiasma" and "chromatid" interference.
The former is the occurrence less often (positive chiasma interference) or more often (negative chiasma interference) than by chance
of two or more exchanges in one bivalent in individual cells in meiosis. The latter is the participation less often (positive chromatid
interference) or more often (negative chromatid interference) than
by chance of the same two chromatids in two, or more, exchanges.

Chiasma interference is an essential feature of Darlington's
(1937) theory of meiosis, which postulates breakage of homologous
chromatids as a consequence of torsional stresses which are released
by breakage, unwinding, and reunion; the unwinding at one point is
supposed to reduce the stress on neighbouring segments of the bivalent so that another break becomes less likely. The fact that in
Aspergillus there seems to be no chiasma interference is hard to
reconcile with that theory.

In *Aspergillus nidulans* Strickland (1958b), in an analysis of 1,231
tetrads from three crosses (one of which is referred to above), found
no chiasma interference between six pairs of adjacent intervals of
sizes ranging from about 5 to 25 units. It is worth keeping in mind
that Aspergillus has much less DNA per map unit than Drosophila,
let alone maize or the mouse. Perhaps chiasma interference is a
feature of genetic structures with a lot of DNA.

Strickland, however, found some evidence, though inconclusive,
of an excess of four-strand double exchanges within individual
marked intervals; in other words an excess of tetrads in which all
four products of meiosis were crossover between two adjacent mark-

ers. Within an interval, an excess of four-strand doubles is the only way in which negative chiasma interference or positive chromatid interference or the occurrence of localised crossing over at the two-strand stage could manifest themselves in tetrad analysis.

We shall see that other types of evidence suggest a quite unexpected feature of crossing over, i.e., the exchanges tend to occur in small clusters. A consequence of this is that within small marked intervals (of less than one unit), the smaller the interval the greater the chance that, if one exchange occurs in it, further exchanges become detectable (Pritchard, 1955). The slight excess of four-strand doubles within intervals found by Strickland could well be a manifestation of the same phenomenon, though diluted by the relatively large size of the intervals.

As to chromatid interference there are at present formidable difficulties in assessing it unequivocally. Strickland's (1958b) results show, *prima facie*, no chromatid interference. Two exchanges, one in each of two adjacent intervals in the same chromosome arm, involve the four chromatids at random. For instance, two-strand, three-strand, and four-strand double exchanges were represented in the 575 tetrads mentioned above in 12, 17, and 11 cases, i.e., in proportions not significantly different from the random expectation of 1:2:1. Other crosses gave similar results.

However, the crude data are known to be affected by the inevitable misclassifications mentioned before. One could try to correct for these, but this implies making groundless assumptions (e.g., that the strand relationships of multiple exchanges remain the same even within small intervals). There seems to be at present no way out. As Sturtevant (1955) put it "a good many of us have become suspicious of detailed quantitative analyses of the phenomenon of interference." These analyses are better put in cold storage until some entirely new approach makes them meaningful. This does not contradict what will be expounded in the section on "The Clustering of Exchanges." The magnitude of the effects described there is enormous: it does not require any detailed quantitative analysis.

The subject of this section has given origin to controversial interpretation. It is one of the most exciting new developments in genetics: the splitting of the gene. According to the definitions given in Chapter II, two recessives of independent mutational origin will be called allelic if the "heterozygote" has a mutant phenotype. If, among the products of meiosis in the "heterozygote"—but not in the two homozygotes—an occasional revertant or an allele different from the two original arises, we conclude that the two alleles are in some way different. We can adopt Roman's (1956) terminology and call the combination "heteroallelic" and the two homozygotes "homoallelic." (Note: We should not call homoallelic a combination of two recessives of different mutational origin from which revertants have not been obtained.)

And now let us go step-by-step through an example of the kind of analysis of heteroallelic combinations which Roper first published in 1950 and which led me to the tentative conclusion (Pontecorvo, 1952) that heteroalleles represent mutations at different sites, distinguishable by recombination within a linearly organised structure, the function of which was, until last year, called a gene. This tentative conclusion has become so well-established since 1950, with evidence coming in from practically every series of alleles investigated with sufficient care in every organism, that Benzer was able in 1956 to propose a new name for the functional unit: the cistron.

In *Aspergillus nidulans* the most extensively analysed is the *ad8* cistron studied by Pritchard (1955, 1958 and unpublished). Seven adenineless independently arisen mutants are located in chromosome I less than one unit distal from *y* and about five units proximal to *bi* (map in Table 5). All seven (*ad20, ad16, ad19, ad12, ad11, ad8, ad10*) are qualitatively indistinguishable on growth response tests, but *ad16* and *ad20* can grow at about one-half the wild-type rate even without adenine. For *ad20* unlinked specific suppressors are known. All seven mutants are completely recessive in haploid

TABLE 10

MAP OF THE *ad8* CISTRON IN *ASPERGILLUS NIDULANS*

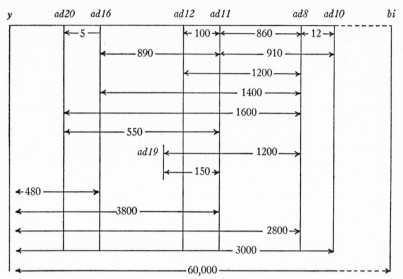

Figures indicate recombination fractions per 10^6. The position of *ad19* relative to *ad12* is not yet established. The order of the mutants is established qualitatively from crosses two-by-two.

Source: Pritchard, 1955, 1958 and unpublished.

heterokaryotic—$(ad) + (ad^+)$—and in diploid heterozygous—ad/ad^+—combinations with wild type. The haploid heterokaryotic and the diploid combinations of any two mutants are adenine-requiring or, in combinations involving *ad16* and *ad20*, intermediate.

Adenine-independent nuclei arise at meiosis from every cross performed (12 out of the 21 possible) between the seven mutants two by two, and mitotically from the three diploid combinations tested ($ad8/ad11$; $ad8/ad16$; $ad8/ad20$). Reverse mutation, as occurring in haploids, cannot account for more than a small proportion of the wild-type nuclei arising in heteroallelic crosses or in heteroallelic diploids.

That the wild-type nuclei from heteroallelic combinations arise from crossing over (meiotic or mitotic) was shown by Pritchard in three ways:

1) The closely linked markers y and bi, 5 units apart astride the $ad8$ cistron, and particularly y which is less than 1 unit away, define the order of any two ad alleles, and this order is consistent over the whole series of seven. For instance (see map, Table 10), of the 365 ad^+ haploid ascospores selected from the cross $\dfrac{y \quad ad8 \quad +}{+ \quad ad11 \quad bi}$, 355 were y and 10 were y^+, suggesting the order y $ad11$ $ad8$ bi. From the cross $\dfrac{y \quad ad8 \quad +}{+ \quad ad16 \quad bi}$, of the 139 ad^+ haploid spores 133 were y and 6 were y^+, suggesting the order y $ad16$ $ad8$ bi. From the cross $\dfrac{y \quad ad11 \quad +}{+ \quad ad16 \quad bi}$, of the 57 ad^+ haploid spores 46 were y and 11 were y^+, suggesting the order y $ad16$ $ad11$ bi. Thus the order of the three ad mutants considered together is y $ad16$ $ad11$ $ad8$ bi. The segregation of the outside marker bi on the other side confirmed the qualitative conclusions based on the more closely linked outside marker y. For some of the heteroallelic combinations the order has also been confirmed by means of mitotic crossing over. For instance, a heterozygote $\dfrac{y \quad + \quad ad8 \quad +}{+ \quad ad16 \quad + \quad bi}$ (or of corresponding arrangement in other heteroallelic combinations) is adenine-requiring. A heterozygote in *trans* of this type should produce a heterozygote in *cis* with arrangement $\dfrac{+ \quad ad16 \quad ad8 \quad +}{y \quad + \quad + \quad bi}$ as a consequence of mitotic crossing over (Chapter V) and segregation of the complementary products to the same pole. This type of segregant is expected to be adenine-independent. A number of adenine-independent segregants were obtained and analysis of some of them by means of haploidization (Chapter VI) showed them to have the segregant genotype indicated above, and in particular to carry both ad mutants on one chromosome.

2) Two mutants (e.g., $ad16$ and $ad8$) in *cis* as just mentioned could be separated by crossing over in crosses to wild type.

3) Crosses carrying three ad mutants, two in one chromosome and one in the homologue (Pritchard, 1958) gave among the ascospores selected for recombination between two of them the appropriate

distribution of markers; e.g., from cross $\dfrac{y \; ad20 \; + \; ad8 \; +}{+ \; + \; ad11 \; + \; bi}$ selection of $ad8^+ \; ad11^+$ ascospores gave 2,696 $ad20$ out of a total of 2,804.

Pritchard's analysis of the $ad8$ cistron thus shows conclusively that crossing over occurs within the cistron, that it gives the two expected reciprocal products, and that these may arise from one (mitotic) event. The complete map of the seven identified sites in the $ad8$ cistron is given in Table 10.

It will be noted that the recombination fractions between the sites are tolerably additive. However, from the data given above it will be noted also that among the recombinants between hetero-alleles in a proportion of cases the outside markers were not reas-sorted. In the three examples given this is most strikingly shown by cross $\dfrac{y \; + \; ad11 \; +}{+ \; ad16 \; + \; bi}$. Out of 57 ad^+ recombinants there were 11 $y^+ \; bi$ and 6 $y \; bi^+$; i.e., in 30 percent of the ad^+ spores the outside markers were not reassorted. This question will be considered in the next section on "The Clustering of Exchanges."

In Aspergillus an acceptably linear arrangement of mutational sites similar to that of the $ad8$ cistron has been obtained in several others, indeed in all the six cistrons so far analysed to a greater or smaller extent (see map, Table 5).

In other organisms with a chromosomal apparatus the best ex-amples are bithorax (Lewis, 1951), lozenge (Green and Green, 1956), and white in Drosophila; and $ad7$ in Schizosaccharomyces pombe (Leupold, 1957): in the latter, nine sites have been located and the proportion of adenine-independent spores from the various hetero-allelic combinations gives a remarkably consistent picture (Table 11). Even though outside markers were not available in this case, the picture is so good that we can do without them.

In organisms with a non-chromosomal system, the outstanding example of mapping in a cistron is that of $rIIA$ in bacteriophage (Benzer, 1955, 1957; Chase and Doermann, 1957) in which over forty sites have been located in unequivocal linear order.

The conclusion of this section is that the analysis of very short map segments leaves very little doubt that crossing over can resolve

TABLE 11

THE *ad7* CISTRON IN *SCHIZOSACCHAROMYCES POMBE*

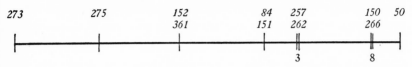

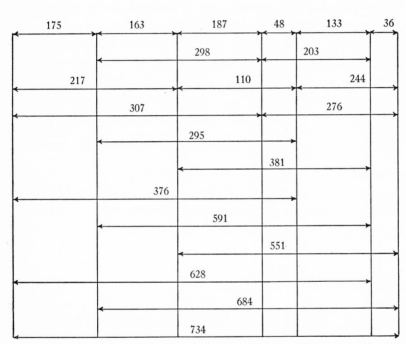

Figures indicate adenine-independent colonies out of 10^6 ascospores from the interallelic crosses shown by the arrows.

Source: Leupold, 1957.

elements of the genome previously assumed to be nonresolvable, i.e., those constituting the structural basis of individual genes (= cistrons). This analysis, furthermore, has revealed certain features of crossing over which are not apparent at more crude levels. These features are the object of the following section.

THE CLUSTERING OF EXCHANGES

In the early stages of the analysis of cistrons in Aspergillus, Roper (unpublished) noted that the bi^+ spores from a heteroallelic cross of two *bi* mutants would have changed the prevalent segregation of the outside marker *y* (five units away) more often than in the expected 5 percent of the cases. Similar observations were made by Roper on the *paba* cistron. In the meantime Giles (1951) observed the same phenomenon in heteroallelic crosses of inositol-less Neurospora. Earlier reports in the literature of what we now understand as possibly the same phenomenon are Sturtevant's (1951*b*) observations on the fourth chromosome of Drosophila.

It was not until Pritchard's (1955) analysis of the *ad8* cistron in Aspergillus, however, that the importance of this puzzling phenomenon was grasped, and vigorously attacked. Besides Pritchard's own work (1955, 1958), the study of the same phenomenon in bacteriophage by Chase and Doermann (1958) and the study in Aspergillus of the effect on map lengths of selecting crossovers in three very short segments by Calef (1957) are very pertinent.

Briefly the situation is as follows. Consider a heterozygote:

Intervals		I	II	III	
	m_1	$+$	s_2		$+$
	$+$		s_1	$+$	m_2
Map units		2	0.1	3	

in which s_1 and s_2 are heteroalleles or very closely linked complementary mutants showing, say, 0.1 percent recombination, and m_1 and m_2 are outside markers also closely linked not more than five units apart. In a random sample of products of meiosis we shall find the following combinations and proportions:

m_1	s_1	s_2	m_2	
$-$	$+$	$-$	$+$	$\}$ Parental 95%
$+$	$-$	$+$	$-$	
$+$	$+$	$-$	$+$	$\}$ Crossovers in I 2%
$-$	$-$	$+$	$-$	
$+$	$-$	$-$	$+$	$\}$ Crossovers in II 0.1%
$-$	$+$	$+$	$-$	
$+$	$-$	$+$	$+$	$\}$ Crossovers in III 3%
$-$	$+$	$-$	$-$	

Double crossovers, in I and III, should constitute less than 1 in 1,000 and double crossovers with one exchange in II, less than 5 in 100,000. If we select the rare crossovers between s_1 and s_2 we expect the outside markers to segregate in one direction in 95 percent of the cases, i.e., 95 percent of these crossovers should be: $-++-$ among the $s_1^+s_2^+$ crossovers or $+--+$ among the $s_1\ s_2$ crossovers; 2 percent should be $+++-$ or $---+$ (crossovers in I and II); and 3 percent should be $+---$ or $-+++$ (crossovers in II and III).

In fact, the situation is dramatically different. Among the selected crossovers we find that the proportion of those showing a further exchange in I or III is very greatly increased above the 2 percent and 3 percent expected. It was shown by Pritchard (1955) that: 1) the additional recombination is the greater the more stringent the selection, i.e., the closer the two heteroalleles crossovers between which are selected; and 2) the effect was highly localised, i.e., the increase in recombination was very large only in the two segments immediately adjacent to that used for selection. The last feature ruled out the possibility that the effect was due to heterogeneity in the population, in respect of proneness to crossing over (Sturtevant, 1955).

An example of the effect of the stringency of selection comes from the comparison of heteroallelic crosses in the $ad8$ cistron. Recombination between $ad8$ and $ad10$ is of the order of 12 per million, and between $ad8$ and $ad12$ of 1,200 per million. In the two crosses of $y\ ad8$ bi^+ by $y^+\ ad10\ bi$ and $y^+\ ad12\ bi$ the distribution of the outside markers y and bi (less than 1 unit and 5 units away, respectively) was:

		I	II	
		y	$ad8$	$+$
		$+$	adx	bi
	$adx = ad10$	Crossover	$adx = ad12$	Crossover
$y\ \ bi$	3	I + II	102	0
$y^+\ bi$	8	II	5	I
$y^+\ bi^+$	14	0	1	I + II
$y\ \ bi^+$	3	I	31	II
	28		139	

Formally, region I was expanded from about 0.003 in nonselective analysis to $6/28 = 0.25$ among $ad8^+$ $ad10^+$ recombinants and to $6/139 = 0.04$ among the $ad12^+$ $ad8^+$ recombinants. Region II was expanded from about 0.05 in nonselective analysis to $11/28 = 0.38$ among $ad8^+$ $ad10^+$ recombinants and to $32/139 = 0.23$ among the $ad12^+$ $ad8^+$ recombinants. This effect of the stringency of selection has been confirmed again and again in Aspergillus and occurs also in phage (Chase and Doermann, 1958).

Calef (1957) showed that it is not a peculiar property of intra-cistron recombination but it occurs also when selecting crossovers between two non-allelic mutants, provided their distance is very small (e.g., $ad9 - paba1$, less than 0.3 units apart).

Pritchard (1955) interpreted these results as follows. The conditions which permit the occurrence of crossing over ("effective pairing") are realised over a few small segments of the chromosome in each cell in meiosis, but distributed along the whole chromosome in a population of cells. Within any one "effectively paired" segment where one exchange occurs, the average number of exchanges is greater than one. The size of the effectively paired segment is small enough to make multiple exchanges detectable only when two markers (allelic or not) are less than about 1 unit apart.

By using triple heteroallelic crosses (Pritchard, 1958) in the $ad8$ cistron, it was possible to calculate that the correlation between exchanges, or "negative interference" as it is also called, extends over at least 0.35 map units. It does not seem to extend, however, over many map units, as Calef (1957) claimed.

Two important questions are: 1) what kinds of relation there are between successive "effective pairing" segments; and 2) how the four chromatids are distributed between successive exchanges within a segment of "effective pairing." This is, of course, a new way of looking at chiasma and chromatid interference.

As to the first, we have no clue other than those from mapping on the usual scale: there may be interference between successive segments, and this would be the basis of classical interference. As to the second, tetrad analysis of intra-cistron recombination has been impracticable so far, but mitotic analysis (Pritchard, 1955, pp. 357–63) shows that two- and three-strand double exchanges do occur. Out of 32 diploid segregants interpreted as resulting from single or multiple mitotic exchanges within the *ad8* cistron, 24 were single. In four of the 8 doubles or multiples, three strands were involved.

One problem should be mentioned here. In order to push the analysis of the fine genetic structure further and further, one has to use closer and closer mutants. We do not know to what extent heterozygosis at one point along a chromosome favours recombination in its immediate vicinity. It is quite possible, therefore, that "negative interference" could be a result of the closeness of the mutants necessary to detect it. If this were so we would have reached in biology a situation of indeterminacy analogous to that well known in physics: beyond a certain limit we could not increase at the same time the accuracy of determination of the relative positions and of the recombination between two markers.

It has been asked in which way the distribution of crossing over in small clusters should affect ordinary mapping. The answer is that provided single exchanges within an "effective pairing" segment are still more frequent than multiple exchanges (and this seems to be the case), the effect would simply be that in ordinary mapping with markers a few units, rather milli-units, apart we underestimate the incidence of crossing over. Maps built on complete cistron analysis would work out at, say, twice the total length that we measure in ordinary mapping. A truer picture of genetic maps will come thus from an extensive use of micro-maps (intra-cistron) and perhaps extrapolation from these to the whole genome, if and when we know that this is legitimate.

UNUSUAL TETRADS

Lindegren (1953) was the first to call attention to the fact that in tetrad analysis occasional tetrads are found in which alleles do not segregate in 1:1 ratio. It is to his great merit that he insisted on the significance of these abnormal tetrads for genetic theory. However, because the examples on which Lindegren based his arguments were disputable in a proportion of cases (Winge and Roberts, 1956; Emerson, 1956; Strickland, 1958), abnormal tetrads were at first not "respectable." In 1955 M. Mitchell produced three indisputable ones from Neurospora. Since then they have become so respectable that speculation has run to the extreme of taking intra-genic recombination as due to the phenomena underlying abnormal tetrads, and ignoring a large amount of evidence to the contrary from Drosophila, Aspergillus and bacteriophage (e.g., M. Mitchell, 1956; St. Lawrence, 1956; Beadle, 1957).

Beadle (1957, page 20) went as far as to state: "In Neurospora, Aspergillus and yeast, intra-genic recombination occurs by a mechanism that can be interpreted as miscopying of small segments of genetic material. This process differs from conventional crossing-over in that a single event does not result in reciprocal products and also by the fact that it does not necessarily lead to a recombination of genetic markers close to and on opposite sides of the gene within which it occurs." Beadle proposed for this hypothetical process the term "transmutation": if a term is necessary at all before we know what goes on, it should be the older one of "gene conversion" adopted by Lindegren.

The two preceding sections have shown, I hope conclusively, that these generalisations may be good enough for particular cases in Neurospora and yeast, and may turn out to be good enough for particular cases in other organisms as well, but they are certainly not applicable to the many accurate intra-cistron analyses so far carried out in Drosophila, Aspergillus, and phage (Chase and Doermann, 1958).

Beadle's statement was clearly based on misconceptions about the following two facts. First, in Drosophila and Aspergillus one intra-cistron recombinational event can, and usually does, give origin to the two reciprocal products. Second, in Aspergillus and phage (perhaps also in Drosophila if preliminary evidence is confirmed), recombination over minute marked segments—i.e., intra-cistron recombination as well as (Calef, 1957) inter-cistron recombination over less than one unit—reveals that exchanges tend to occur in clusters:

the smaller the marked interval the greater the probability that if one exchange occurs in it further exchanges nearby become detectable. This leads to the fact that a high proportion of recombinants within a small interval go with additional exchanges immediately outside it, hence the unexpected distribution of the outside markers. Generalising from one well-investigated case in Neurospora (the *pyridoxineless* mutants studied by M. Mitchell) and from a few others not so well investigated, is like generalising from the case of *Bar* in Drosophila to all cases of mutation or recombination.

Having cleared the air of misconceptions, let us now see something about abnormal tetrads and what they tell us. I shall refer exclusively to the work of Strickland (1958*a*) in *Aspergillus nidulans* because it is the one which has involved the maximum number of marked loci with linkage relations unquestionably established: 1,642 fully classifiable tetrads from nine crosses involving fifteen loci. Among these 1,642 tetrads there were 17 abnormal ones, i.e., tetrads in which one or more pairs of alleles were present in ratios other than 1:1. Of these 17, 11 could well be due to inevitable technical limitations, as mentioned before; the other 6, or at least 4 of them, were unlikely to be due either to faults in technique or to situations easily explainable on classical theory. These 6 are described in detail in Table 12.

In three cases (asci 8, 9, 10) there is a 3:1 ratio (3 bi^+/1 bi) at a particular locus; i.e., reverse mutation or "conversion." In one case (ascus 12) all the 8 loci, one of which is on a different chromosome from the other 7, show abnormal ratios of 5:3 or 6:2: this can be most easily interpreted as a case in which one of the four products of meiosis degenerated and the vacancy of the corresponding two nuclei of the third division was filled by two other third division nuclei dividing again. Ascus 14 could also be explained in the same way or by the occurrence of a process as that which has given origin to ascus 13 (discussed in the following paragraph). As only three different products of meiosis were recovered from ascus 14 it is not possible to decide between these two alternatives.

Ascus 13 is the most remarkable of all, and it constitutes an ines-

TABLE 12

ASCI, WITH SEGREGATION RATIOS OTHER THAN 1:1, NOT LIKELY TO BE DUE TO FAULTY TECHNIQUE OR TO GENETICAL SITUATIONS EXPLAINABLE ON CLASSICAL THEORY

Ascus number	Genotype of zygote*	Total No. of classifiable asci from this cross	Genotypes of spores of abnormal ascus	No. of spores germinating	Abnormality	Suggested "explanation"
8	pro1 + + + bi1 / + paba1 y ad8 +	392	+ paba y ad + pro + + + bi pro + + + bi	4 2 2	3:1 bi+/bi	conversion
9	pro1 + + + bi1 / + paba1 y ad8 +	392	+ paba y ad + pro paba y ad + pro + + + bi pro + + + +	2 1 2 1	3:1 bi+/bi	conversion
11	ribo + ad14 + paba1 y + / + an + pro1 + + bi1 pyro4	264	+ an + + paba y + bi pyro ribo + + ad pro + + bi + pyro ribo + + ad pro + + bi + pyro	4 2 2	3:1 bi+/bi	conversion
12	ribo + ad14 + paba1 y + / + an + pro1 + + bi1 pyro4	264	ribo + ad + paba y + bi + ribo + ad + paba y + bi + + an + pro + + + bi pyro	3 2 3	All alleles in abnormal ratios	extra mitotic division
13	+ ad17 + + bi / pro1 + paba1 y +	151	pro + paba + bi ad + + y + ad paba y + + ad + + y bi	1 2 1 2	3:1 pro+ad/ pro ad+ crossover in minute segment ad-paba	conversion over at least 8 units
14	+ ad17 + bi / pro1 + paba1 y +	151	pro + paba y + bi pro + paba + bi + ad + y +	2 3 2	3:1 pro ad+ paba/pro+ad paba+ crossover in paba-y	extra mitotic division or conversion over at least 8 units

Total of these 3 crosses — 807

Total of other 6 crosses which did not give asci with unquestionably abnormal ratios — 835

1,642

* Maps of Aspergillus nidulans are in Table 15. Source: Strickland, 1958a.

capable example and extension of the reality of a process which M. Mitchell (1955) postulated as possible for individual loci. The postulate was that in the replication of a heterozygous locus the defective part is not replicated and the rest is replicated twice. Ascus 13 shows that more than this can happen: a long segment of chromosome including two markers—the mutant allele *ad17* and the wild-type allele *pro$^+$*, 8 units apart—was replicated twice. This segment, which occurs in triplicate among the four products of meiosis, is immediately proximal to a position of exchange within 0.3 units. This is an example of "conversion" of a chromosome segment at least 8 units long. The absence of other markers leaves it undecided if in fact the "conversion" affected the whole chromosome segment "left" of the position of exchange (Table 13).

TABLE 13

ABERRANT ASCUS 13 SHOWING CROSSING OVER IN INTERVAL *ad17-paba1*, SECOND CROSSOVER IN ADJACENT INTERVAL *paba1-y* AND DOUBLE REPLICATION OF A LARGE SEGMENT

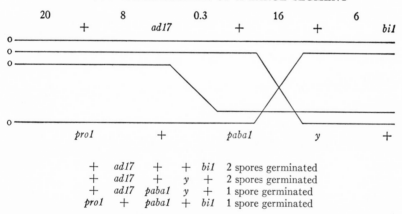

+	ad17	+	+	bil	2 spores germinated
+	ad17	+	y	+	2 spores germinated
+	ad17	paba1	y	+	1 spore germinated
pro1	+	paba1	+	bil	1 spore germinated

The segment *proximal* to the exchange between *ad17* and *paba1* replicated twice. This segment is marked over 8 units: the absence of other proximal markers, or markers on the other arm, makes it impossible to know whether or not this extra replication involved the whole chromosome to the left of the position of exchange or only the marked 8 units.

Source: Strickland, 1958a.

In total, excluding ascus 12 which is most likely an example of differential multiplication of the products of meiosis, we are left with three cases of "conversion" at the *bi* locus; and one certain case (ascus 13), and another possible case (ascus 14), of conversion of the same long segment, starting very near to (ascus 13, and possibly ascus 14) a position of exchange.

The association between exchange and conversion which M. Mitchell first noted for an individual locus is certainly evident here, but differently from her hypothesis, the misreplication extends over long chromosome segments.

We can now summarise Strickland's finding of unquestionably abnormal segregations. The 1,642 fully classifiable tetrads (i.e., with three of four products of meiosis recovered) from nine crosses and involving 15 loci, can be tabulated in order to see in how many asci each locus has shown abnormal segregation out of all the asci in which it could have shown it.

The data are as follows (after the symbol of each allele the fraction indicates the number of asci with abnormal ratios for that allele over all the fully classifiable asci heterozygous for that allele):

ad1 0/134; *ad8* 0/392; *ad14* 0/336; *ad15* 0/575; *ad17* 2/151; *an* 0/264; *bi1* 3/1419; *met1* 0/48; *paba1* 1/1489; *pro1* 2/1489; *pro3* 0/575; *pyro4* 0/435; *ribo* 0/264; *s12* 0/88; *w3* 0/179; *y* 0/1446.

"Conversion," or double replication of part of one homologue, seems to be a rare phenomenon in general, but not so rare with particular alleles, e.g., *bi1*, and/or associated with crossing over in particular regions, e.g., the minute segment *ad17-paba1*. It will be noted that *ad8* (the allele which gives the name to the extensively mapped *ad8* cistron) gave no abnormal segregations in 392 asci: in terms of single-strand analysis this represents $392 \times 4 = 1,568$ ascospores.

The next step in this field is, of course, the development of techniques for analysing selectively the rare abnormal tetrads, without having to dissect thousands or tens of thousands of normal ones.

Another step is that of using more extensively "half-tetrads" recoverable from mitotic crossing over, i.e., the technique developed by Roper and Pritchard (1955), and applied by Pritchard (1955) to Aspergillus and by Roman (1956) to yeast. The cistrons analysed in this way in Aspergillus (*ad8* and *paba1*) showed that mitotic recombination between alleles is usually reciprocal, though this technique could not exclude, of course, that "conversion" occurred in a minority of cases. The cistron analysed in this way in yeast by Roman (1956), on the other hand, has shown—at least as far as published— only "conversion."

In view of the characteristic behaviour of the individual cases we need a wide collection of fully analysed examples before we can hope to understand what are the peculiar features of those cases—a small minority—which show "conversion." The analogy with the story of *Bar* comes immediately to mind: it would have been disastrous for progress if a general working model had been based on it.

REPLICATION AND RECOMBINATION

Any working model for the mechanism of intra-chromosome recombination in organisms with standard meiosis has to satisfy three essential requirements: 1) recombination is reciprocal; 2) each event involves only two out of the four products of meiosis; and 3) double or multiple crossovers may involve all four, or only three, or only two, of the four chromatids. I should be inclined to include as a fourth essential requirement to be satisfied that crossing over occurs in small clusters (with an average of less than two exchanges per cluster), but this is too new a finding to insist on it.

There has long been a feeling (going as far back as Belling, 1931) that a working model of recombination which is part and parcel of the mechanism of chromosome replication would be more useful than the model of breakage and reunion, which was brought to such a plausible refinement by Darlington (1937).

The reasons why the breakage-reunion model is no longer satisfactory are: 1) the difficulty of translating it into physico-chemical processes open to experiment; 2) the fact that one of its postulates (that breakage at homologous positions is a consequence of replication of paired and relationally coiled homologues) is contradicted by the timing of DNA and chromosome replication (e.g., Taylor 1957); and 3) the difficulty of using it as a general model valid also for recombination in bacteriophage, in bacterial transformation, and in bacterial transduction. Two further objections could be the occurrence of mitotic crossing over, and the fact that crossing over occurs in small clusters. Neither of these two is insuperable.

The trouble is that no alternative, satisfactory for organisms with standard meiosis, has been suggested. In bacteriophage there seems to be a substantial case for a model of "copying-choice" (Lederberg, 1955; Levinthal, 1954). During the template-copying process of replication of one genetic structure the "copy," which is made running along one template, would switch over and be continued along its homologue. In its simplest form this model is inapplicable to recombination in meiosis because it permits recombination between only two chromatids—the "new" ones.

It has been claimed, without any grounds, that phage recombination is a nonreciprocal process. Even if it were nonreciprocal, this is not an essential feature of a copying-choice model. The timing of replication in phage could be such that the switch has a very small probability of occurring reciprocally at precisely the same point. But this need not be a general case. We can easily imagine a timing such that the switch of copy A' from template A to template B at position x, makes it inevitable for copy B' to switch from template B to template A at the same position.

If nonreciprocity is not a necessary feature of copying-choice models of recombination, they would be compatible with one of the requirements of meiotic crossing over, i.e., the formation of the reciprocal products of one event.

That chromosomes may reproduce by a template-copying mechanism is strongly suggested by the remarkable work with isotopes (Plaut and Mazia, 1956; Taylor, 1957). The disagreement between the two groups of researchers is about the timing and the precise mode of replication rather than about the essential fact that the labelled atoms of a parental chromosome are not distributed at random between either the daughter chromosomes (Plaut and Mazia) or the grand-daughter chromosomes (Taylor).

Thus chromosome replication is, to use the terminology of Delbrück and Stent (1957), either "conservative" or "semi-conservative," or at least partially conservative.

If the replication of the chromosomes is by a template-copying process, and a model of recombination based on switch in copying-choice is not incompatible with the production of reciprocal recombinants, the only requirement still to be satisfied is that all four products of meiosis must be involved in recombination.

A possible, but not elegant, way out is that proposed by Lindegren and Lindegren (1937) and Schwartz (1955). A breakage-reunion process between "old" and "new" structures, i.e., between "sister chromatids" in classical terminology, is invoked for distributing among the four products of meiosis the recombination which took place by copy-choice only between "new" structures.

The difficulty with this model, apart from the inelegance of assuming two different processes of exchange, is that nobody as yet has been able to devise critical experiments for the classic problem of whether or not "sister chromatid" exchanges do occur. But chromosome breakage is something easy to produce, as radiation genetics shows. Breakage of old and new structure at precisely homologous positions should present no problem when the new structure is in the exact alignment with the old one, an alignment that is implicit in the fact that the new structure has just been produced. All considered, there is something useful in a two-stage model like this. The use of isotope-labelled chromosomes may well offer the critical experiment necessary for deciding whether or not sister chromatid exchanges do occur. The results of Taylor (1957), suggestive as they are, do not exclude as yet that the sister chromatid exchanges observed are, in fact, breakage-reunion effects due to the disintegrations of the label.

Now let us consider a further question: what different problems are set by three basically different kinds of template-copying model? One is of the kind proposed by Haldane (1954) and Stent (1958) in which, reduced to the essentials, a template produces a complementary template and the latter produces two like the original. Calling the complementary templates positive and negative:

POSITIVE ⟶ NEGATIVE ⟶ 2 POSITIVE
(the original POSITIVE may be scrapped)

Another one is of the kind in which the template itself is made up of two complementary parts, like the Watson and Crick structure for DNA and the Kacser's protein-DNA structure. In this case replication is by "complement formation" (Delbrück and Stent, 1957):

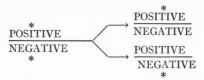

(* parental parts of the complementary templates)

A third kind of template-copying model is that suggested by Penrose and Penrose (1957 and unpublished) in which the template is a symmetrical structure which can replicate either by accretion on one side or on both:

$$\overset{*\,*}{AA} \to \overset{*}{A}\overset{}{A} + \overset{*}{A}\overset{}{A} \text{ or } \overset{*\,*}{AA} \to \overset{*\,*}{AA} + AA$$

Models of the first type offer no difficulty for recombination. The difficulty is to give them a precise testable form, as Stent (1958) has attempted to do. Clearly, if two positives are made simultaneously along each of two homologous negatives, any two of the four positives could switch over from one negative to the other. The trouble with this model is that there is no preservation at all of the molecular integrity of the parent structures as required by the Plaut and Mazia, and Taylor experiments.

Models of the complement formation type have one essential difficulty: the switch of copying-choice can occur only between congruous members of the complementary structures. If we symbolise each of two homologous complementary structures as $\overset{*\,*}{WC}$ and $\overset{*\,*}{W'C'}$, replication will give $\underset{1}{\overset{*}{W}C} + \underset{2}{W\overset{*}{C}}$ and $\underset{3}{\overset{*}{W'}C} + \underset{4}{W'\overset{*}{C'}}$ and switch of copying-choice could only occur between 1 and 3 or 2 and

4. In terms of chromosomes, a system of replica by complement formation permits the occurrence of 2- and 4-strand multiple exchanges but not of 3-strand multiple exchanges. It agrees, however, with the results of Plaut and Mazia, and even more with those of Taylor, which suggest some sort of preservation of the parental molecular integrity in chromosome replication.

The Penrose model has not yet been elaborated in detail, especially as to the lengthwise bonds of the structure, but it offers an important possibility. If in different segments along the replicating chromosome the process of replication may be either semiconservative ($\overset{*\,*}{AA} \rightarrow \overset{*}{A}A + A\overset{*}{A}$) or conservative ($\overset{*\,*}{AA} \rightarrow \overset{*\,*}{AA} + AA$), "sister chromatid" crossing over becomes only a change from one to the other type of replication. This, however, raises the old questions of classical genetics as to the "plane of splitting" of chromosomes.

As things stand there does not seem to be any easy way of reconciling the essential results of intra-chromosomal recombination in meiosis with the physico-chemical models available. A two-step system with a complement formation mechanism for replication, and with a copying-choice plus sister-chromatid crossing-over mechanism for recombination, or a system of the type proposed by Stent, come nearer to what is required.

I should like to suggest that all these models should be re-examined to see whether the following addition would help, i.e., that the chromosomal elementary fibril (in Ris's 1957 sense) is double, i.e., made up of two identical templates. This might meet one of Stent's (1958) postulates that "mating" of two identical structures is necessary for replication.

Finally, let us consider the question, raised repeatedly, of the basic unity or otherwise of the processes of linked recombination at various levels of genetic complexity. There are three possibilities: 1) recombination as observed in phage, or as resulting from transformation and transduction, on the one hand, is basically different from that in meiosis; 2) the two systems are basically identical; 3) intra-cistron recombination in meiosis is basically identical to that in phage etc., but inter-cistron recombination in meiosis is based on a different process.

I hope to have shown that there are no grounds for the last proposition. We have to face squarely the first and second alternatives. My preference is for the second. Let us hope that work in progress will permit us soon to decide.

CHAPTER V

MAPPING CHROMOSOMES VIA MITOTIC RECOMBINATION

THAT segregation may occur regularly though rarely in hetero-zygous somatic cells of Drosophila is unquestionable since the classic work of Stern (1936). The occurrence of somatic crossing over (or "mitotic" crossing over because it occurs also in cells of organisms without "soma"), or at least of some process of segregation in somatic or vegetative cells, is also inferred in a number of higher organisms including, perhaps, man (Goudie, 1957). In microorganisms it is unquestionable in *Aspergillus nidulans*, *Aspergillus niger*, *Aspergillus sojae*, *Aspergillus oryzae*, and *Penicillium chrysogenum* (see Chapter VI), and probable in *Fusarium oxysporum forma Pisi* (Buxton, 1956) and yeast (James and Lee-Whiting, 1955; Roman, 1956).

In all these the analysis and the inference have been purely genetic. There are innumerable cytological reports of mitotic chiasmata, of course, but it is not sure what the observed configurations really were. As yet, not one has linked the genetic results of mitotic crossing over with cytological observations, as it has been done for meiotic crossing over. The idea, hard to die, that somatic pairing as cytologically detectable at metaphase in a few organisms has something to do with mitotic crossing over is not very helpful to say the least, though Boss's (1955) review suggests that it might have some value. As a result of the work with *Aspergillus nidulans* (summary: Pontecorvo and Käfer, 1956, 1958) genetic analysis of mitotic crossing over is now so advanced that the complete blank on the cytological side is far from disastrous.

In the experiments of Stern (1936), Drosophila heterozygous in *cis* for the linked recessives *y* (yellow) and *sn* (singed), and therefore with grey body colour and straight *setae*, showed occasional spots mainly either yellow or yellow singed. The spots were of variable size, but mainly including only one or a few *setae* and a small patch of cuticle.

Flies heterozygous for *y* and *sn* in *trans* gave mainly either yellow or singed spots sometimes isolated but more often next to each other, i.e., twin spots in which a yellow small area was contiguous to a singed area, both of course in the overwhelming background of wild-type tissue. Flies heterozygous for *y* and *sn* in *cis* gave no twin spots.

With Drosophila, Stern could not, of course, isolate cells from one of these spots, grow them into a whole fly, and then analyse the genotype of the fly through sexual reproduction as we can do with Aspergillus. Yet he was able to deduce through a brilliant series of experiments that a large proportion of these segregant spots could be accounted for by mitotic crossing over at the four-strand stage followed by the normal mitotic type of segregation of sister centromeres: Stern's diagram of mitotic crossing over is in the textbooks.

Stern could not deduce then, as we can now, what proportion of the segregant spots could also arise in another way, i.e., by haploidisation, though undoubtedly the bulk of the segregant spots in his flies could not have had this origin. In fact, it would be most instructive to reanalyse today his impressive lot of data, keeping in mind this additional possibility.

MITOTIC CROSSING OVER IN *Aspergillus nidulans*

In an organism like *Aspergillus nidulans*, isolated single vegetative cells can form a colony which goes through the complete life-cycle, including the sexual cycle. As in most other filamentous fungi, in *Aspergillus nidulans* the nuclei of the vegetative cells are usually haploid, and the diploid stage is limited to just the zygote which immediately goes through meiosis.

Mitotic crossing over obviously requires vegetative cells with diploid nuclei, and these were produced by Roper (1952) in a very simple way. Since then it became apparent that what was produced artificially in the laboratory is probably a process which plays an important part in the genetic system of this and other moulds in nature. In fact, in moulds lacking a sexual cycle, the formation of diploid nuclei in vegetative cells is probably one part of a cycle which substitutes sexual reproduction as the basis of genetic recombination (Chapter VI).

The summary which follows is based on research by various workers at the Department of Genetics, Glasgow University. A recent monograph (Pontecorvo and Käfer, 1958) gives the up-to-date picture and all the references. For general reference to symbols and techniques in the genetics of *Aspergillus nidulans* the reader is referred to Pontecorvo *et al.* (1953) and Käfer (1958).

The analysis of mitotic recombination in *Aspergillus nidulans* and similar moulds begins with diploids heterozygous for a number of known markers distributed between the two sets of parental chromosomes. It is advisable, of course, to use the same markers in different arrangements between the homologues. For instance, in the work of Pontecorvo and Käfer (1956, 1958) the two diploids used had the same relevant nine markers (plus a few others), but they were differently distributed between two chromosome pairs, which were thus subdivided into ten analysable intervals (Table 16).

Mitotic crossing over is a rather rare event—it occurs at the rate of, say, one exchange every few hundred nuclei: multiple exchanges can be disregarded in first approximation. Its effect is to produce from a heterozygous nucleus two daughter nuclei each homozygous for either one or the other homologous segment distal to the position of crossing over (Table 14).

Mitotic crossing over being rare, we cannot take a random sample of all nuclei and classify them as to parental or segregant by the phenotypes of hyphae carrying them in homokaryotic condition. We must select homokaryotic hyphae carrying crossover nuclei.

To select, we require suitable markers which lend themselves to

TABLE 14

DIAGRAM OF THE CONSEQUENCES OF MITOTIC CROSSING OVER IN
AN ARM OF A MULTIPLY-HETEROZYGOUS CHROMOSOME PAIR

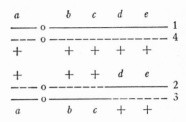

A. *Segregation of both crossover chromatids (2 + 3) to the same pole*

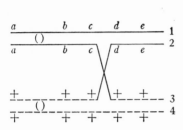

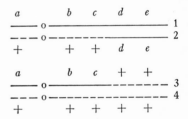

B. *Segregation of one crossover and one non-crossover chromatid to the same pole*

Segregation A does not lead to any phenotypic change, except in the case of hetero-allelic *trans* diploids where crossing over within the cistron leads to the *cis* arrangement.

Segregation B yields daughter nuclei homozygous for the segment distal to the position of crossing over.

selection. This is the decisive factor in efficiency. So far we have used five types of selection, but many more are easily designed:

1) *visual selection*, i.e., for markers, such as colour of the conidia, which can be detected in a very small group of cells. This is of course the only type of selection which Stern could use in his work with Drosophila.

2) *selection for recessive suppressors:* if a diploid strain is homo-zygous for a recessive determining a nutritional requirement, and heterozygous for a recessive "suppressor" of this requirement, homozygotes for the suppressor will be automatically selected on a

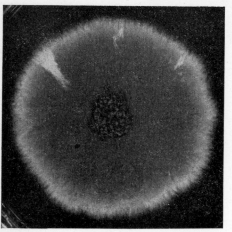

A

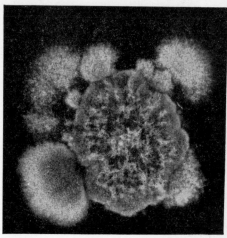

B

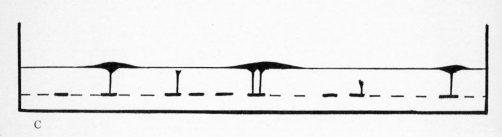

C

THREE METHODS FOR THE IDENTIFICATION AND
ISOLATION IN *ASPERGILLUS NIDULANS* OF MITOTIC
SEGREGANTS FROM HETEROZYGOUS DIPLOIDS

A) Spots showing the recessive colour yellow (y/y or haploid y) appear over the sporulating surface of an otherwise green colony, heterozygous $y/+$.

B) On medium containing acriflavine in concentrations permitting only their stunted growth, heterozygous semiresistant colonies ($Acr/+$) throw out vigorous sectors fully resistant because homozygous Acr/Acr or haploid Acr.

C) On medium devoid of adenine (bottom layer) colonies homozygous $ad20/ad20$ and heterozygous for a recessive "suppressor" $su-ad20$, grow as a very thin mycelium. After growth for 48 hours the colonies are marked and another layer of the same medium is poured above them. Hyphae homozygous su/su, or haploid su, grow quickly through the top layer and establish vigorous colonies above it.

medium containing inadequate amounts of the relevant growth factor.

3) *selection for recessive or semidominant resistance to harmful agents:* if a diploid is heterozygous for one such marker, on a medium containing the harmful agent the homozygotes, with higher resistance, will be selected.

4) *selection by starvation:* it is known that on media lacking the relevant growth factors germinating conidia of Aspergillus strains having certain combinations of two requirements survive longer than those of strains having one only of these. For instance, on a medium lacking adenine and biotin, double adenineless-biotinless mutants survive much longer than biotinless ones. Thus homozygous adenineless-biotinless segregants are easily selected from a diploid homozygous biotinless but heterozygous for adenine requirement, and therefore adenine-independent.

5) *selection of* cis *from* trans *heterozygotes:* a strain heteroallelic in *trans* for a nutritional requirement (e.g., carrying in *trans* two allelic pyridoxineless recessive mutants) will not grow on a medium lacking the relevant growth factor. The corresponding *cis* heterozygote, resulting from mitotic crossing over between the two sites, will not require the growth factor, and can thus be selected.

Methods of selection 1), 2), and 3) are illustrated in the plate.

The next point in mitotic crossing over analysis is that the markers used for selection of homozygotes must be as distal from the centromere as possible. The nearer they are to the tips of the chromosome, the better; and we need one selector so located for any arm which we wish to analyse. This is not an easy requirement, of course, but in view of the ease with which one can obtain "suppressors" and mutants conferring resistance to poisons, it is not as difficult as it looks.

The reason for this requirement is clear. A mitotic exchange, followed by the appropriate type of segregation (Table 14), leads to homozygosis for the whole chromosome segment distal to the position of exchange. Hence a selector marker located near to the centromere would give little information because any homozygote for it would also be homozygous for all the other markers in coupling with it on the same arm. It could tell us that these other markers are linked with the selector and distal to it, but nothing about their order.

The essential of mitotic crossing over analysis is this: among the homozygotes for a distal marker some will be also homozygotes at the next proximal locus (i.e., nearer to the centromere), some at the next two, some at the next three, etc. Homozygotes for any one, two, three, etc., segments nearer to the centromere than the selected marker, expressed as a fraction of all homozygotes for the selected marker, give the proportional incidence of exchanges in each of the corresponding intervals between the centromere and the first marker, the centromere and the second, etc., up to the selected marker. This reasoning has similarities with that used in the meiotic analysis of attached-Xs in Drosophila.

We can also locate the centromere, in a chromosome marked on both arms, by identifying the segment where homozygosis for one of two adjacent markers always goes with homozygosis for a series of other markers in one direction, and homozygosis for the other marker always goes with homozygosis for a series of markers in the other direction. Suppose we have the order $a\,b\,c\,d\,e\,f\,g\,h$ and we find that homozygotes for e are always homozygous for $f\,g\,h$ but not for $a\,b\,c\,d$, and homozygotes for d are always homozygous for $a\,b\,c$ but not for $e\,f\,g\,h$. We would conclude that the centromere is between d and e.

The procedure in mitotic analysis is thus to select, from a heterozygous diploid, segregants homo- or hemizygous for certain distal markers and then to analyse these segregants for their residual genotypes.

When the markers linked with the one used for selection are recessive, and in *cis* with it, the position of crossing over is immediately deduced from the phenotype. For instance: if d is the selector in $\dfrac{a\ \ b\ \ c\ \ d}{+\ +\ +\ +}$, crossovers in the interval $c\text{-}d$ will be only d in phenotype, crossovers in the interval b-c will be $c\ d$ in phenotype, crossovers in the interval a-b will be $b\ c\ d$ in phenotype, and crossovers between the centromere and a will be $a\ b\ c\ d$ in phenotype.

But if some or all of the proximal recessive markers are in *trans* $\left(\text{e.g., } \dfrac{+\ +\ +\ d}{a\ b\ c\ +}\right)$ then all crossovers between the centromere and d will be phenotypically alike, i.e., $a^+\ b^+\ c^+$ and the difference in the position of exchange results only in genotypic difference as between homo- and heterozygosis for these dominant alleles. This difference can only be ascertained by further genetic analysis, either meiotic or mitotic.

HAPLOIDISATION AND MITOTIC MAPPING

At this point it is necessary to introduce another process, which was not conspicuous, though probably present, in Stern's work with Drosophila, but which is both biologically and technically very important in moulds. I refer to haploidisation.

A proportion of between one-tenth and one-half of the phenotypically recessive patches from heterozygous diploid colonies work out to be haploid (Pontecorvo, Tarr-Gloor, and Forbes, 1954). If we use diploids multiply marked on several chromosomes, it works out that the haploids originating from them are of all the possible genotypes expected from *recombination between non-homologous chromosomes without crossing over*, i.e., linked markers segregate almost always together.

On the basis of perhaps one thousand haploid segregants classified so far, we can quite certainly conclude that mitotic crossing over and haploidisation are the result of different processes, occurring in different nuclei and coinciding in one nucleus not more often than would be expected by chance. Provisionally we may conclude that they are independent phenomena.

Chapter VI will deal in detail with haploidisation as a biologically important process. Here it is only necessary to stress what a convenient tool it is in genetic analysis. As it yields all possible recombinations between chromosomes, but practically no recombination between linked markers, it locates at once any unlocated marker in its appropriate linkage group.

If we isolate, it does not matter with which selective technique, segregant spots from heterozygous colonies, some will be haploid and some diploid. For an analysis of mitotic crossing over clearly we have to separate the two types. But later we can also use haploidisation of diploid segregants to analyse further the genotypes of these segregants.

There are various ways of sorting out haploid from diploid segregants. One is the measurement of the diameter of the conidia which are, in *Aspergillus nidulans*, spherical uninucleate cells, the volume of which in diploids is double that of haploids. Other methods are, as it were, automatic. It is always advisable to use a combination of methods.

Let me exemplify one of these automatic devices, of which there are many. In a diploid $o\dfrac{pro \quad paba \; + \quad + \quad w}{+ \quad + \quad y \quad Acr \; +}o$ yellow segregants (haploid y, diploid y/y) will be hemizygous Acr, i.e., fully resistant, if haploid; and will be heterozygous $Acr/+$, i.e., half resistant, if diploid; they cannot be hemizygous sensitive (Acr^+) because in this case they would also be hemizygous w, which is epistatic to y, and therefore they could not be yellow (see Table 5 for symbols and locations). The haploids are easily identified by testing for acriflavine resistance.

Once the diploid segregants for any one marker have been sorted out from all the segregants, we have to classify them as to their residual genotypes. This is done mainly by isolating further haploid segregants from them.

Generalising, if we have two chromosomes marked with selectors $(s'$ and $s'')$ and other markers (m', m'') : $\dfrac{s' \quad + \; + \quad s''}{+ \quad m' \; m'' \quad +}$ we first isolate homozygous s'' ("first order" segregants). These will be still heterozygous $\dfrac{s'}{+}$ and may be of any one of the three following constitutions as to the multiply-marked chromosome:

$$\frac{+ + \; s''}{m' \; m'' \; s''} \begin{matrix}(1)\\(2)\end{matrix} \qquad \frac{+ + \; s''}{m' \; + \; s''} \begin{matrix}(1)\\(2)\end{matrix} \qquad \frac{+ + \; s''}{+ \; + \; s''} \begin{matrix}(1)\\(2)\end{matrix}$$

according to where the exchange which gave origin to any one of them occurred. If now we search for s' "second order" segregants from one of these "first order" segregants, and of these "second order" segregants we consider only the haploids, among the haploids the two s'' chromosomes—(1) and (2)—will be represented in about equal numbers, and the genotype of the diploid "first order" segregant will be ascertained by the types of haploid "second order" segregants recovered from it.

An example of complete mitotic analysis of two multiply-marked diploids is given in Table 16. This mitotic mapping included two linkage groups and a total of ten analysable regions. As extensive mapping based on crosses had already located all the markers used for mitotic analysis, it is possible to compare meiotic and mitotic maps (Table 15).

The absolute incidence of mitotic crossing over is small, difficult to measure, and probably variable. Thus, we cannot build maps based on mitotic recombination fractions. We can instead build maps in which the distances are expressed in different units in each chromosome arm. This is done by taking the total segregants for the distal selected marker and expressing the incidence of crossing over in each of the intervals between the centromere and that marker as fractions of that total.

In order to compare this relative incidence of mitotic crossing over in each arm with that of meiotic crossing over, we express the latter in fractions, for each interval, of the map distance of that interval out of the total map distance from centromere to distal marker.

The distribution of mitotic crossing over, as compared to that of meiotic crossing over, has so far revealed no definite trend. In the right arm of chromosome I the maximum concentration of mitotic crossing over is in a 7-unit segment about 20 units from the centromere. In the left arm, the maximum concentration of mitotic crossing over is within 20 units from the centromere, but lack of further

TABLE 15

COMPARISON OF MAPS OF THE CHROMOSOMES I AND II OF *ASPERGILLUS NIDULANS* BASED ON MEIOTIC AND ON MITOTIC CROSSING OVER

Units / Relevant regions (linkage maps):

Chromosome I:
su —39— ribo —19— an —7— ad14 —20— pro —20— paba —8— y —16— ad20 —0.2— bi (6)
Intervals: I (su–ribo), II (ribo–an), III (an–ad14), IV (ad14–pro), V, VI, VII

Chromosome II:
Acr —25— w —21— ad1 —50
Relevant regions: a —0— b

Crossovers as percent of total crossovers in each arm

Intervals	__ Chromosome I __ Left arm (85 units)					Right arm (50 units)				Chromosome II (45 units)		
	I	II	III	IV	Total	V	VI	VII	Total	a	b	Total
Mitotic — Diploid Y	23	8	6	69	100	6	72	22	100	13	87	100
Mitotic — Diploid Z	24	6	8	64	100	5	73	22	100	18	82	100
Mean	23	8	—	69	100	6	72	22	100	15	85	100
Meiotic	45	23	8	24	100	41	20	39	100	57	43	100

(For the Meiotic row, a brace groups intervals III + IV = 32.)

Diploid marker constitutions

Diploid Y

Chromosome I: →su + ribo an + + + pro + paba + + ad20 bi
Chromosome II: →Acr w +

Diploid Z

Chromosome I: su← + ribo an + + + pro + →y← + ad20 bi
Chromosome II: →Acr w← +

The percentages above are based on 405 crossovers for I Left, 293 for I Right, and 502 for II. The results from the two diploids were made equal to 100 both in meiotic and mitotic analysis and the intervals expressed as fractions of this total. The total length of each arm has been made equal to 100. The arrows indicate the markers used for selection.

Source: Pontecorvo and Käfer, 1956.

MITOTIC SEGREGANTS FROM A DIPLOID OF THE FOLLOWING CONSTITUTION IN RESPECT OF CHROMOSOME I

Constitution of the diploid:

su-ad20	ribo	an	+	+	+	ad20	bi
+	+	+	pro	paba	y	ad20	+

Meiotic units: su-ad20 –39– ribo –19– an –27– ● (centromere) –20– pro –8– paba –16– y –0.2– ad20 –6– bi

---

SELECTION A: YELLOW SEGREGANTS (y or y/y)

HAPLOIDS

su	ribo	an	pro	paba	y	ad20	bi	
+	+	+	pro	paba	y	ad20	+	126
others								0
								126

DIPLOIDS

su	ribo	an	pro	paba	y	ad20	bi	
su	ribo	an	+	+	y	ad20	+	35
+	+	+	pro	paba	y	ad20	+	
su	ribo	an	+	paba	y	ad20	+	110
+	+	+	pro	paba	y	ad20	+	
su	ribo	an	pro	paba	y	ad20	+	9
+	+	+	pro	paba	y	ad20	+	
								154*

SELECTION B: ADENINE-INDEPENDENT SEGREGANTS (su or su/su)

HAPLOIDS

su	ribo	an	pro	paba	y	ad20	bi	
su	ribo	an	+	+	+	ad20	bi	80
others								0
								80

DIPLOIDS

su	ribo	an	pro	paba	y	ad20	bi	
su	ribo	an	+	+	+	ad20	bi	59
su	ribo	+	pro	paba	y	ad20	+	
su	ribo	an	+	+	+	ad20	bi	21
su	ribo	an	pro	paba	y	ad20	+	
su	ribo	an	+	+	+	ad20	bi	181
su	+	+	pro	paba	y	ad20	+	
								261

---

* In addition 9 "nondisjunctional" diploids homozygous for all the markers on both arms of the chromosome carrying y.

The types of the diploid segregants establish the sequences centromere — pro — paba — y — ad20 — bi for the "right" arm (Selection A) — and centromere — an — ribo — su-ad20 for the "left" arm (Selection B). The haploids from Selection B show that the two sequences represent two arms of the same linkage group.

The full genotype of each of the diploid segregants was not ascertained in every case. The markers used for selection (in bold type) were su on the left arm and y on the right arm.

Source: Data of Pontecorvo and Käfer, 1958.

markers does not permit us to be more precise. In the left arm of chromosome II again this maximum concentration is within 18 units from the centromere, and preliminary evidence suggests that the same is true for the right arm.

Clearly, we cannot yet generalise but there does not seem to be the overwhelming concentration of mitotic crossing over very near the centromere which Whittinghill (1955) found for spermatogonial and oögonial crossing over induced by X-rays in Drosophila.

It is necessary here to stress again the following incidental point. That mitotic crossing over in Aspergillus gives origin to reciprocal products of exchange is unquestionable: it has been ascertained both by isolating and testing the genotypes of both members of what corresponds to a "twin spot" in Drosophila and, more crucial, by recovering the two complementary products of one exchange within one nucleus (Roper and Pritchard, 1955). The recovery of the complementary products in one nucleus is based on $2+3 \div 1+4$ segregation (see Table 14).

CONCLUSIONS

The most important result of these investigations is that mapping by means of mitotic recombination is not only feasible, but easy. Indeed, because of the rarity of multiple exchanges in mitotic crossing over and the occurrence of haploidisation without crossing over, it is qualitatively more dependable than meiotic mapping.

Mitotic recombination has already been used for mapping in organisms in which the sexual cycle does not occur like *Aspergillus niger* (Pontecorvo, Roper and Forbes, 1953; Hutchinson, 1958), *Aspergillus sojae* (Ishitani *et al.*, 1957), *Penicillium chrysogenum* (Sermonti, 1955). It can be used, of course, for "breeding" of industrial strains with improved qualities.

There is no reason in principle, though there is probably great technical difficulty, why it should not be used for the formal genetics of higher organisms, including man, in tissue or organ cultures, as suggested by Pontecorvo and Käfer (1954; 1956; 1958). Even if mitotic crossing over did not occur in higher organisms, processes akin to "haploidisation" certainly occur, at least as rare accidents of mitosis. The approaches illustrated here show that even with this limitation the scope of genetic analysis of tissue cells would be enormous.

Finally, nothing suggests any difference in basic mechanisms between meiotic and mitotic crossing over. This makes mitotic crossing over a convenient tool, in some cases technically more convenient than meiotic crossing over, for the study of crossing over in general.

And now I would like to offer some evolutionary speculation. In organisms in which the life cycle is practically all in the diploid stage, mitotic crossing over has to be kept in check because it leads to mosaic soma. We can assume that in higher organisms there are mechanisms keeping its rate as low as possible. Perhaps these mechanisms have developed *pari passu* with the evolution of the diploid stage.

Examples of organisms in which crossing over is suppressed at meiosis are known and not uncommon: for instance it is prevented at gametogenesis in the Drosophila male and in the silkworm female. Furthermore, meiotic crossing over is variable under a series of genetic and environmental factors. It is conceivable that similar mechanisms are available for preventing crossing over at mitosis.

All this brings us back to the speculations outlined in the last chapter: crossing over is a primitive feature, perhaps inevitably connected with the physical chemistry of chromosome duplication, and therefore occurring perhaps between sister chromatids in haploids. To lead to genetically relevant results it must occur between homologues, i.e., it requires a diploid stage no matter how transient. In the evolution of the diploid stage it has first been utilized for recombination, perhaps in transient diploids as found in phages and bacteria today, and finally, as the differentiation between soma and germ track became sharper, it has been prevented in the soma but not in gametogenesis where its timing and correlation with haploidisation have become quite precise.

CHAPTER VI

NOVEL GENETIC SYSTEMS

IF ONLY one outstanding contribution of microbial genetics to biological thinking had to be singled out, it would be this: the realisation that transfer of genetic information from one individual, or cell,
to another is not the monopoly of sexual reproduction. Hence the
first subtitle of this chapter borrowed from an article by Haldane
(1955).

ALTERNATIVES TO SEX

Sexual reproduction may be defined as the regular alternation in
the life cycle of an organism of karyogamy and meiosis. Its main
biological significance lies in the fact that it achieves recombination,
in one line of descent, of genes derived from different lines of descent.
With the diploid or polyploid condition which goes with it in all
higher organisms, sexual reproduction stores an enormous amount
of genetic variation under phenotypic uniformity. It is a remarkable
means of pooling the genetic information of the individuals of an
interbreeding group.

The efficiency of sexual reproduction in terms of population fit-
ness for present conditions and flexibility for the future is impressive
(e.g., Mather, 1943), so much so that biologists looking at evolution
through neo-Darwinian field glasses have tacitly ignored in the past
the inconvenient exceptions—e.g., the huge number of species of
microorganisms in which sexual reproduction was and is not known
to occur or goes without an extended diploid stage. Without sexual
reproduction the pooling of genetic information was supposed to be

impossible, and without the diploid stage, permitting heterozygosis, the "storage" of a large variety of genetic information was supposed to be impossible.

In short, the "genetic systems" (Darlington, 1939, 1956) of the myriads of asexual microorganisms did not make sense. We understand now that this conclusion was wrong only because it was based on the assumption that sexual reproduction was the sole means of transfer and storage of genetic information.

The idea that the bacteria and the fungi imperfecti could get along in the struggle for survival merely by means of mutation within clones has never been very convincing. The reasons are vividly explained in Muller's lecture "The Dance of the Genes" (1947b). So has the idea that, e.g., homothallic fungi use sexual reproduction within a system of complete self-fertilization. Equally unconvincing is the contention that a clonal system of rapid multiplication compensates for the single-cell life cycles of most microorganisms and for their haploid condition by permitting a faster and more ruthless operation of natural selection.

It is to the credit of microbial genetics to have shown that this scepticism was justified. The different kinds of mechanism for the transfer of genetic information and for its storage which have come to light in microorganisms already number at least five. Apart from the pioneer work of Griffith (1928) on transformation in pneumococci, they have all been discovered since 1945.

We know now of at least four novel systems in bacteria: transformation (review: Hotchkiss, 1955), bacteriophage-mediated transduction (Zinder and Lederberg, 1952), quasi-sexual reproduction à la Lederberg (1947), and lysogenisation (Lwoff, 1953; Jacob and Wollman, 1957); and of at least one in fungi (Pontecorvo, 1954), which is the main subject of the present chapter. All that these systems have in common with sexual reproduction is that they bring together in one cell lineage hereditary determinants from separate cell lineages: in other words they ensure "genetic recombination."

I have proposed (Pontecorvo, 1954) that processes of genetic recombination, other than a regular alternation of karyogamy and meiosis as in sexual reproduction, should be called "parasexual." The etymology of this is that they lead to the same end but by a different way. Lederberg (1955), and Lederberg and Lederberg (1956), however, have proposed a different terminology, and Wollman, Jacob and Hayes (1956) have proposed a third. All three suggestions have some good points and I shall attempt to unify them as follows.

The Lederbergs and Wollman *et al.* based their suggestions on the consideration of whether only a part of the genome of one cell or the whole of it is transmitted to another cell to form a zygote. This will be a partial zygote ("merozygote," in Wollman *et al.*'s terminology) in the first case, or a complete zygote in the second. The complementary term "holozygote" is the obvious one for the latter.

Lederberg proposed a few years ago the general term "transduction" for processes leading to the formation of what we now call a merozygote and defined it as "the transmission of a (nuclear) genetic fragment from a donor cell (which in every case so far is destroyed in the process) to a recipient cell which remains intact." He gave as example the two already known categories of transduction in bacteria which are distinguished by the agency of transmission: i.e., DNA, as in Pneumococcus and Hemophilus transformations; or a phage particle, as in Salmonella and *Escherichia coli* transductions *sensu strictu.*

Furthermore, Lederberg defined sex as "equivalent to karyogamy, the formation of a hybrid zygote from the fusion of two intact 'gamete' nuclei." Thus, in the above terminology, sex (or should it read "sexual reproduction"?) is the process which gives origin to a holozygote, transduction is that which gives origin to a merozygote.

Now, a system of genetic recombination does not consist only in the transfer of genetic material, but also in what follows this transfer. The term "parasexuality" takes care of the latter aspect. It was suggested for a process occurring in *Aspergillus nidulans* (Chapter

V), side by side with an orthodox sexual cycle. This process leads to recombination by means of mitotic crossing over and "haploidisation."

In a species where both sexual and parasexual recombination occur side by side, it is easy to see what the difference is between the two. Both the sexual and the parasexual cycles in *Aspergillus nidulans* start with the formation of a holozygote. It is irrelevant for the distinction that in the sexual cycle the holozygote is formed by the fusion of two haploid nuclei in highly specialised (sexual) organs, and in the parasexual cycle it is formed erratically by presumably random fusion of two haploid nuclei in nonspecialised multinucleate cells. What is relevant is that in the sexual cycle the holozygote, or cells descended from it, undergoes meiosis, i.e., a series of events integrated to the minutest detail involving recombination, segregation, and reduction to the haploid condition. In contrast, in the parasexual cycle, the holozygote, and cells descended from it, do not undergo meiosis; they undergo recombination and reduction with no timed coordination; in fact, the two processes take place independently in different nuclei of a lineage at different times, as will be shown later.

We therefore have to consider two features of any system of genetic recombination: one is the mode of formation of the zygote—and here the suggestions and terminology by Lederberg and Wollman *et al.* are very acceptable. The other is the mode of recombination, segregation, and reduction.

Accordingly, there are the following four possible combinations of these two features, and for each combination examples are given:

		Made of formation of the zygote	
		Transfer of whole genome (holozygotic)	*Transfer of a part of the genome (merozygotic)*
Made of recombination, segregation, and reduction	*Coordinated (e.g., meiotic)*	1) sexual cycle in Ascomycetes and higher organisms	3) transduction in *E. coli*, transformation in Pneumococcus
	Uncoordinated (e.g., mitotic)	2) parasexual cycle in filamentous fungi	4) lysogeny in *E. coli* (?)

Whether or not transduction and transformation are correctly classified under 3) rather than under 4) and lysogeny under 4) will have to await more precise knowledge of what goes on after a DNA fragment, or a phage-carried fragment, have entered the recipient cell. The classification of the system of conjugation in *Escherichia coli* is also controversial: it may be properly classified under 1) in the case of certain strains and under 3) or even 4) in the case of other strains.

Of the systems just mentioned, the one in fungi provides a complete *ersatz* for sexual reproduction, i.e., it provides for both recombination and sheltering of gene variation. In the latter respect, in fact, it is more versatile than sexual reproduction because it includes two ways of storing gene variation instead of only one, i.e., heterokaryosis in addition to heterozygosis.

The systems in bacteria are as yet too imperfectly analysed to know how and to what extent they provide for the sheltering of variation. The guess is that they will turn out to go a long way.

One thing is certain: in microorganisms a good deal of sheltering is provided by intercellular associations of genetically different cells. Syntrophism is the best known example of this: two types of cells, nutritionally exacting in different ways, can supplement one another when growing next to one another. This is only one degree removed from the intracellular complementarity of two kinds of nuclei within a common cytoplasm in heterokaryons, and two degrees removed from the intranuclear complementarity of two sets of chromosomes carrying non-allelic mutants in heterozygotes. The general impression gathered from the five types of novel systems of recombination discovered so far is one of great versatility—a striking contrast with sexual reproduction.

The elements of sexual reproduction are extremely uniform. True, meiosis may be partially suppressed in certain species or in one sex, but where fully developed it is basically of one type. It is only in the secondary paraphernalia of sexual reproduction that the variety of nature is perhaps at its best.

By contrast, in these other novel systems we range from a virus acting as vector of genetic materials to a system of accidental karyogamy and accidental haploidisation not timed with crossing over. The surprises which we have had in these first ten years are such that we should be prepared for more to come.

THE PARASEXUAL CYCLE IN FUNGI

And now it is time to consider in some detail the parasexual cycle in filamentous fungi, with which I have firsthand acquaintance. Like all four other novel systems of recombination, it has been identified and analysed in the laboratory but we know nothing as yet about how widespread is its occurrence in natural populations.

The parasexual cycle was first identified in *Aspergillus nidulans* (Pontecorvo and Roper, 1952) where it exists side by side with the sexual cycle. It was then identified in *Aspergillus niger* (Pontecorvo, 1953; Pontecorvo, Roper, and Forbes, 1953) where there is no sexual cycle. Soon after it was found in *Penicillium chrysogenum* (Pontecorvo and Sermonti, 1954), and then in *Fusarium oxysporum f. pisi* (Buxton, 1956) and in *Aspergillus sojae* (Ishitani et al., 1957). The complete cycle is now unquestionably known in *Aspergillus nidulans, Aspergillus niger* (Hutchinson, 1958), and *Penicillum chrysogenum*, but in the other species mentioned one or more of the steps await confirmation. The details of the cycle have been worked out to a considerable extent (Pontecorvo, Tarr-Gloor, and Forbes, 1954; Pontecorvo and Käfer, 1956, 1958), and a review was published recently (Pontecorvo, 1956).

The complete cycle consists of the following steps: 1) Fusion of two unlike haploid nuclei in a heterokaryon to give a diploid heterozygous nucleus (Roper, 1952). This nucleus may go on multiplying side by side with the haploid nuclei in heterokaryotic hyphae, but may eventually be sorted out to produce heterozygous diploid homokaryotic hyphae; 2) Occasional mitotic crossing over during the multiplication of the diploid nuclei; 3) Occasional haploidisation of the diploid nuclei.

We can assume that in heterokaryotic hyphae (and in homokaryotic hyphae as well) fusion is not restricted to unlike nuclei, but occurs equally between like nuclei. In this case the result would be only the formation of homozygous diploid nuclei reverting eventually to haploid: i.e., there would be no genetically relevant effect. This is possibly what E. Sansome (1949) observed long ago in *Penicillium notatum*.

If, on the other hand, the nuclei which fused were genetically different, the diploid nucleus would be heterozygous, mitotic crossing over in it could lead to segregation and recombination of linked genes, and haploidisation would lead to recombination of genes on different chromosomes.

Haploid nuclei of unlike genotype are found together in heterokaryotic mycelium usually as a consequence of previous anastomosis between hyphae of different origin. Hence the parasexual cycle in fungi is genetically significant only with heterokaryosis, just as the sexual cycle is significant only with heterozygosis. Heterokaryosis is an integral part of the parasexual cycle. The whole cycle is given diagrammatically in Table 17.

The preceding chapter gives the full details of mitotic crossing over. Here it will be necessary to give only details of haploidisation. Let us take again one of the examples from Pontecorvo and Käfer (1958) given in Table 16. One of the diploids used was of green colour (because heterozygous $y/+$ and $w/+$), adenine-requirer (because homozygous *ad20* and heterozygous for the recessive suppressor *su-ad20*), and partially resistant to acriflavine (because heterozygous $Acr/+$). Its full genotype was as follows:

Chromosome 1 $\dfrac{su\text{-}ad20 \quad ribo \quad an}{+ \qquad\quad + \qquad +}$ o $\dfrac{+ \quad + \quad + \quad + \quad bi}{pro \quad paba \quad y \quad ad20 \quad +}$

Chromosome 2 $\dfrac{Acr \quad w}{+ \quad\;\; +}$ o $\dfrac{ad1}{+}$

Chromosome 5 $\dfrac{pyro}{+}$

Chromosome 8 $\dfrac{chol}{+}$

Visual selection of yellow segregants (y/y $w/+$ diploids or y w^+

TABLE 17

STEPS IN THE PARASEXUAL CYCLE

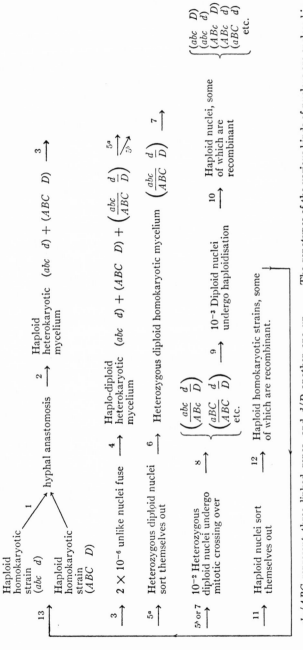

abc/*ABC* represent three linked genes, and *d*/*D* another one unlinked. The genotypes of the various kinds of nucleus are enclosed in parentheses.

haploids) from this diploid gave 163 diploids and 126 haploids. (This is an unusually high proportion of haploids. More often they represent some 10 to 20 percent of all segregants. A high proportion, however, is useful to illustrate the present point.)

In respect of chromosome 2, the 126 haploids were all w^+ and acriflavine sensitive and, with one exception, they were all $ad1$. The exception was $ad1^+$, i.e., probably an example of coincidence, or subsequent occurrence, of crossing over and haploidisation in the same nucleus or in the same nuclear lineage. That in respect of chromosome 2 the haploids, selected as phenotypically yellow, should carry the w^+ member of the pair is only a consequence of epistasis of w over y: those carrying the w member would have been white, not yellow, and therefore not included in the sample. The important point is that 125 out of 126 haploids carried all three alleles of the w^+ strand, i.e., only one crossing over had occurred in 126 on a length of at least 80 meiotic units.

In respect of chromosome 1, all 126 haploids, selected for the y member of the pair, carried all the other alleles on both arms of that member, i.e., su^+, $ribo^+$, an^+, pro, $paba$, and bi^+: again, no crossing over in a length of over 130 meiotic units.

This example is one of the many showing that mitotic crossing over is not associated with haploidisation. It cannot show that segregation of whole, non-homologous chromosomes is random because in both cases one member of each pair was selected. But now let us consider the two other marked chromosomes. In their respect the 126 yellow haploids were distributed as follows:

Parental associations			*Recombinant associations*		
pyro	*chol*	24	+	*chol*	40
+	+	35	*pyro*	+	27
		59			67

There were the four possible types, with parental and recombinant associations compatible with random segregation. This behaviour has been confirmed again and again, though in different cases the allele ratios, i.e., the proportions of recovery of the two homol-

ogous strands may depart from 1:1 because of differential multiplication, as they do in the above example in respect of the $pyro/+$
segregation (51 $pyro$: 75 $pyro^+$). We must keep in mind that the
segregants are recovered many nuclear generations after the event
which produced them.

That the process of haploidisation might be the consequence of
accidental and rare (say 1×10^{-3}) failure of regular mitotic separation of sister chromatids is suggested by two facts: 1) Most haploids
can be shown to originate as aneuploids ($2n - a$, where $a = 1$ at the
onset of the process and $a = n$ at the end). They carry at first some
chromosomes in disomic condition, which are rapidly reduced to the
monosomic condition by accidental loss and selection in favour of
the fully balanced haploid (Pontecorvo and Käfer, 1958; Käfer,
1958)(Table 18); 2) Among the diploid segregants a proportion—

TABLE 18

HYPOTHETICAL STEPS IN THE PROCESS OF HAPLOIDISATION

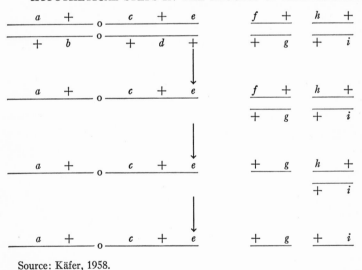

Source: Käfer, 1958.

say 5 percent—are homozygous for all the markers on both arms of
one chromosome. It is true that these segregants could originate as
a consequence of two-strand double exchanges across the centro-

mere, but this would require an enormous "coincidence." It seems more likely that they represent one of the consequences of the same process of the breakdown of mitosis which gives origin to haploids: i.e., "non-disjunction" in respect of one chromosome pair of the set.

THE PARASEXUAL CYCLE IN THE GENETIC SYSTEM

Now we are in a position to assess the part played by the parasexual cycle as a way of achieving gene recombination and of storing gene variation in the genetic system of the sexual Ascomycete *Aspergillus nidulans.*

In very rough figures, based on the use of induced markers and laboratory conditions, we can take the following:

Proportion of heterozygous nuclei in a freshly
synthesized heterokaryon   : 10^{-6}
Proportion of haploid nuclei produced by
diploid nuclei   : 10^{-3}
Proportion of crossover nuclei produced by
diploid nuclei   : 10^{-2}

Without selection the equilibrium between diploid and haploid nuclei in a population would be $10^{-3}/10^{-6} = 1{:}1000$. This would be true in populations made up exclusively of heterokaryotic hyphae: in fact, isolates of filamentous fungi from nature are known to be often, but not always, heterokaryotic (Jinks, 1952). Thus haploid strains should grossly outnumber diploid strains in nature, unless there were a strong selection in favour of the latter.

The second point to consider is the amount of recombination possible with the parasexual cycle. The following calculations are, I repeat, very crude. Only 10^{-3} of all nuclei in a mycelium or a population of mycelia are heterozygous, and recombination in them occurs at the rate of one mitotic crossing over in 10^2 nuclei. In these 10^{-3} nuclei, furthermore, random recombination between the 8 chromosome pairs occurs once in 10^3. We can thus calculate the "recombination index" per nucleus, adapting to mitotic recombination Darlington's (1937) formula for meiotic recombination (which requires a minor correction). Let us use the following symbols:

No. of exchanges per diploid nucleus   E
No. of chromosome pairs   n
Proportion of diploid nuclei per colony   d
Proportion of diploid nuclei per colony which
    undergo haploidisation   h
Mitotic recombination index: $I_{mi} = [E + (n - 1)h]d$
Substituting the values mentioned above:

$$I_{mi} = [10^{-2} + (8 - 1) \, 10^{-3}] \, 10^{-3} = \text{about } 2 \times 10^{-5}$$

Let us now compare this index with that from meiotic recombination in a heterokaryotic mycelium or a population of mycelia. The proportion of nuclei going through meiosis out of all nuclei in a fully fertile colony can be estimated at about one in a thousand. Thus the recombination index via meiosis per nucleus of a colony can be calculated as follows:

No. of chiasmata per nucleus in meiosis   C
Proportion of nuclei going through meiosis   m
Meiotic recombination index:

$$I_{me} = (C + n - 1) \, m$$

Substituting:

$$I_{me} = (13 + 8 - 1) \, 10^{-3} = \text{about } 2 \times 10^{-2}$$

This value will have to be halved because not more than 50 percent of the asci are of crossed origin. Thus the amount of recombination occurring through the parasexual cycle is about 500 times smaller than that through the sexual cycle.

It looks as if in a species like *Aspergillus nidulans*, in which the sexual cycle is still present, the parasexual cycle could contribute only to a limited extent to gene recombination (always assuming that there is no selection in favour of diploid nuclei).

On the other hand, in the two asexual species of which we have experience (*Aspergillus niger* and *Penicillium chrysogenum*) the parasexual cycle seems to be more active than in *Aspergillus nidulans*. Diploidisation occurs at a higher rate, and recombination, at least via mitotic crossing over, seems to be more frequent. It is easy to see that a small shift in the equilibrium in favour of diploids, and a small increase in the rate of mitotic recombination, can make the contribution of the parasexual cycle as important as that of the sexual cycle in terms of the amount of recombination it yields. It can therefore replace the sexual cycle where it has been lost.

And now for the other aspect of a system based on the parasexual cycle—the sheltering of gene variation. This is the consequence of dominance and heterozygosis in higher organisms with an extended diploid stage. Filamentous fungi are more versatile: they have both heterokaryosis and heterozygosis for this purpose.

But there is more, heterokaryosis is a physiologically more pliable system than is heterozygosis. In a heterozygote there are only three possible genotypes in respect of a pair of alleles: A/A; A/a; and a/a: i.e., 100 percent of A, or 50 percent of A, or 0 percent of A. In a heterokaryon all possible nuclear ratios (and therefore allele ratios) can occur between 0 and 100 percent of A.

It is known, furthermore, that heterokaryosis is also a system of vegetative adaptation, as the nuclear ratios can be changed during growth as an adaptive response to external conditions (Beadle and Coonradt, 1944; Pontecorvo, 1946; Ryan and Lederberg, 1946; Jinks, 1952).

We can picture a population, i.e., a colony of filamentous fungus or a group of spatially non-isolated colonies as something very dynamic and quite different from what we were led to believe they were until a few years ago. The colony is far from limited to mutation in a haploid clone as its only means of genetic variation. It can receive and absorb immigrant nuclei and cytoplasm from neighbouring colonies, it can try out all sorts of combinations and ratios of different nuclei for vegetative adaptation, it can shelter recessives both in heterokaryotic condition and in heterozygotic condition, and it can try out all sorts of combinations of genes both in diploid and haploid condition.

The population genetics of systems based on the parasexual cycle is not likely to be dull. Unfortunately the methods of population genetics are not in my line. I can only hope that somebody with the necessary competence will become interested in parasexual genetic systems from this angle. Most of the fungi employed in industrial fermentations and most of those pathogenic to cultivated plants are asexual. Some at least are already known to have the parasexual cycle as a substitute. Apart from the fundamental interest of studying the population genetics of parasexual organisms, there should be enough demand from applied genetics for the sort of approach just mentioned.

CONCLUSIONS

THE six chapters of this book discuss in detail the present trends in certain basic fields of genetics. It may be useful now to summarise. Some mistaken ideas are on the way out; others—not necessarily correct—have appeared either as replacements or as novelties.

THE GENETIC CODE

The most obvious wrong idea on its way out is that of the particulate gene, i.e., of the genetic material as beads on a string in which each bead is an ultimate unit of crossing over, of mutation and of specific activity. This picture was not merely crude: it was wrong because it implied an unnecessary, and almost certainly nonexistent, structural differentiation between the beads and the string. Against it Goldschmidt has fought for years and Muller stressed long ago that it was not called for either on theoretical or on experimental grounds. What has replaced it is the picture of a nonrepetitive linear sequence of building blocks of only a few different kinds, the unique groupings of which determine unique functions. Each of these functions we now call a "cistron."

The analogy of the genetic material with a written message is a useful commonplace. The important change is that we now think of the message as being in handwritten English rather than in Chinese. The words are no longer units of structure, of function, and of copying, like the ideographic Chinese characters, but only units of function emerging from characteristic groupings of linearly arranged letters. Miscopying has now become misspelling: a mistake in letters or in their order, not usually a mistake in words. In this analogy, letters correspond to mutational sites exchangeable by crossing over, words correspond to cistrons, and misspellings to mutations. When the confusion of the present transitional period will be over and when knowledge about the primary functions of the genetic material will be sounder, we may be able to use again the term gene without danger, both for the group of mutational sites, the function of which is to determine the amino-acid sequence of a protein, and for the function itself. This is what I shall do in the following paragraphs.

It will have to be seen whether or not—continuing with the analogy—the words usually follow one another without overlaps. If so they may join by links not different from those between letters in cursive, or they may follow one another with the interposition of meaningless short sequences. They might also be linked by bonds of a different nature from those between letters. But in this case either all crossing over would have to be intra-genic or there would be two different kinds of crossing over—inter- and intra-genic—a possibility for which there is no evidence, to say the least.

The picture of the genetic material as a continuous nonrepetitive sequence is the result of the refinement of genetic analysis. I gave it shape some years ago though most of its experimental basis came later, mainly from work with Aspergillus, bacteriophage, and Salmonella.

While this picture was emerging from genetic analysis, equally promising became the search for what, in terms of physical chemistry, the letters and words of the genetic message are. First, from work with the transforming principles of Pneumococcus, then from bacteriophage, and finally from tobacco mosaic virus, it became clear that genetic information could be coded—and actually is coded in these organisms—in either desoxyribo- or ribo-nucleic acids. It will have to be seen if in respect of DNA this is the case in all organisms or only in what Stanier calls the "lower protista," which have DNA not conjugated with histones.

Parallel with the realisation that genetic information could reside in nucleic acids came the brilliant contribution of Watson and Crick. The complementary duplex structure which they proposed for

DNA, and which seems to be essentially correct, offers just what is required for the genetic material: 1) it is an aperiodic structure with infinite possibilities for coding information; 2) it suggests a new model for replication, i.e., complement formation by each of the two parts of a duplex structure; 3) it is capable of carrying the kind of information which is necessary for specifying the 20 aminoacids found in proteins.

The convergence of the structural analysis of DNA with the biological analysis of the genetic material is one of the most exciting events in biology. It is only fair to remember, however, that Astbury was the first to suggest, twenty years ago, some of the biological implications of the structure of DNA, which was then just beginning to be formulated.

Even if the Watson-Crick ideas turned out to be wrong in respect of the mechanism of replication, they would have had a most fruitful effect: they have led to formulate more clearly what are the problems of replication of the genetic material, and of determination of phenotypic differences. They have also led to a series of most elegant experiments and we shall soon know if it is possible to homologise the map of the mutational sites of a gene (perhaps in terms of nucleotide pairs) with the sequence of aminoacid residues of the protein, the specificity of which is determined by that gene.

GENE ACTION

These advances, both in ideas and experiments, have made the one gene-one enzyme hypothesis of Beadle more plausible now, in some modified form, than it was ten years ago.

While it seems likely that part of the genetic message consists of specifying individual proteins, a question for the future will be whether this is the only way in which the genetic material works. This is the same as asking whether the only information in the genetic message is in the individual words and none emerges from the arrangement of words and groups of words.

The existence of position effects shows, of course, that supragenic arrangement is relevant. This evidence, however, does not help to discriminate between relevance at the code level or relevance at levels removed from it. In other words, the question is whether irrespective of their location two genes determine the same two proteins (which may work differently according to where they happen to be synthesized) or the two genes, if differently located, determine different proteins.

Another very big question, related to the preceding one, is that of the amount and nature of the genetic information required for morphogenesis.

It would be useful in this respect to estimate the number of different proteins in one or more organisms with little or no morphogenesis, e.g., bacteria and moulds. If this number were of the same order as the estimated number of cistrons there would be a strong suggestion that in these organisms the bulk of the genetic material has but one function: one gene-one protein. Then the basic question in the genetics of morphogenesis would become that of whether or not the genes operating on morphogenetic processes also act by nothing more than determining the specificity of proteins. In the affirmative, morphogenesis would result from the epigenetic evocation of protein-forming systems, with gene-determined information for each protein. In the opposite case, we would have to search the genetic material for information of a nature other than that of specifying aminoacid sequences.

It would not be surprising if a part of the genetic control of morphogenesis were of the first kind. But it would be surprising if it were all, and the "higher fields" mentioned in Chapter III had no reality.

It may be advisable to start looking at the problems of morphogenesis from the angle of information theory: it could well be as rewarding as it has been in the case of the genetic determination of protein specificity.

REPLICATION AND RECOMBINATION

The genetic material has to be replicated every time a cell divides, and has to recombine, typically at meiosis. While it is clear now of what replication consists, recombination is still as baffling a problem as it ever was.

In replication, an existing structure, made up of a unique linear sequence of a large number of building blocks of a few different kinds, determines the selection of an identical sequence out of an almost infinite number possible for the arrangement of building blocks from a random pool.

Whether or not the existing sequence is a duplex polynucleotide replicating by complement formation or, as Stent suggests, by transfer of information to an RNA-protein intermediate, or is something as yet not described, is a matter of detail: the essentials of replication can hardly be different from those indicated above.

The study of recombination suffers from the same disease as Italian literature: that of having reached the highest peak too early in its life. The result is subsequent scholasticism. The *Divina Commedia* of recombination was Darlington's *Recent Advances in Cytology*.

After thirty years of scholastic development, recombination has to be approached anew with a completely fresh mind. What is needed is to start with information other than that from microscopic observation, which is too coarse.

The facts that have to be kept in mind are those from genetics, especially tetrad analysis and analysis of minute regions, those from electronmicroscopy, and those from the correlation of meiotic stages with the time of replication of the chromosomal materials.

There are three main questions: 1) What are the basic molecular processes underlying recombination and what relations do they bear to those of replication? 2) Are these molecular processes fundamentally the same from phage to man? 3) If they are the same, what accounts for the great differences in their end results between various organisms?

Among the important new facts to be kept in mind are mitotic crossing over and the clustering of exchanges. They force us to reconsider the dogma that cytologically visible pairing, as observed at zygotene and later, is a condition for crossing over. Discontinuous contacts extending over minute homologous regions, and possibly occurring well before zygotene, seem to be a better model.

This model reduces the enormous differences in incidence of crossing over between organisms, or between sexes of one organism, or between meiosis and mitosis, merely to differences in the incidence of these contacts. Furthermore, this model is compatible both with the occurrence of crossing over well before meiotic prophase and with its coincidence with the time of duplication of the chromosomal material, known to be completed before zygotene. Cytologically visible pairing along the whole length of homologous chromosomes— a mechanical device essential for segregation—may well turn out to have nothing to do with crossing over.

The last three years have witnessed a promising beginning in the electronmicroscopy of the chromosomes, and in the use of autoradiography to follow the fate of their atoms in replication.

One of the results of electronmicroscopy I should like to single out: what Ris calls the elementary chromosome fibril appears to be about 100 Å in diameter and made up of two fibres of nucleoprotein about 50 Å in diameter lying side by side. When deproteinised each of these fibres is about 20 Å in diameter.

If confirmed, this structure would have two interesting features: one is its symmetry across the longitudinal axis; the other is that it would offer the means of satisfying, within one chromosome, the requirement for mating of identical structures which Stent suggests to be necessary for replication.

Clearly, decisive advances in the electronmicroscopy of chromosomes are badly needed to fill the appalling gap between molecular structure and microscopic structure.

There is a widespread expectation that crossing over will turn out to be based not on mechanical breaks and reunions but on some sort of switch in copying-choice as part of the replication process. The use of autoradiography with meiotic chromosomes, so successfully used by Mazia and Plaut and Taylor with mitotic chromosomes, may show how well grounded is this expectation.

New techniques but, above all, an unprejudiced mind, are necessary for a start towards an understanding of the most basic of all genetic phenomena: recombination.

THE VERSATILITY OF RECOMBINATION

An unexpected development of the last fifteen years is the discovery that sexual reproduction is not the only process of recombination: transduction, transformation, cytoplasmic infection, mitotic recombination, autogamy, are all recent additions.

"Merozygotic" processes, i.e., processes of transfer of only part of the genome, are now taken for granted. This means, incidentally, that the narrow view that genetics is nothing but the study of the modes of formation of the gametes can no longer be entertained even formally. Its holders could still cling to it, however, by stretching the term gamete to any vehicle of genetic information from one cell to another.

In respect of evolution theory, the discovery of the versatility of recombination has made nonsense of the specious arguments which used to be produced for reconciling with neo-Darwinism the widespread occurrence of asexual microorganisms. We realise now that if in an organism there is no obvious sexual cycle, we had better find out which other process of recombination is operating. But perhaps the most promising outcome of these advances is the fact that they have opened up the field of the genetics of somatic cells.

Two obvious approaches come immediately to mind. One is the study of differentiation as a process involving transformation-like or transduction-like transfer of genetic information from cell to cell.

The other, for which I am responsible, is the analysis of the genotype of a donor by means of mitotic segregation in cultures of its cells. This kind of analysis, which bypasses sexual reproduction, may soon well make the knowledge of the genetics of man and of other slow-breeding organisms more extensive than that of Drosophila.

WORKS CITED

Ambrose, E. J., and A. R. Gopal-Ayengar. 1953. Molecular orientation and chromosome breakage. Heredity (Suppl.). 6: 277–92.

Aschaffenburg, R., and J. Drewry. 1955. Occurrence of different beta-globulins in cow's milk. Nature. London. 176: 218–19.

Avery, O. T., C. M. MacLeod, and M. McCarthy. 1944. Studies on the chemical nature of the substance inducing transformation of pneumococcal types. J. Exp. Med. 79: 137–58.

Barish, N., and S. A. Fox. 1956. Immuno-genetic studies of pseudoallelism in *Drosophila melanogaster*: II Antigenic effects of the vermilion pseudoalleles. Genetics. 41: 45–57.

Barratt, W. R., D. Newmeyer, D. D. Perkins, and L. Garnjobst. 1954. Map construction in *Neurospora crassa*. Adv. Genet. 6: 1–93.

Beadle, G. W. 1945. Biochemical genetics. Chem. Rev. 37: 15–96.

—— 1957. The role of the nucleus in heredity, in The Chemical Basis of Heredity, pp. 3–22. Baltimore, Johns Hopkins Press.

Beadle, G. W., and V. L. Coonradt. 1944. Heterocaryons in *Neurospora crassa*. Genetics. 29: 291–308.

Beadle, G. W., and S. Emerson. 1935. Further studies of crossing over in attached-X chromosomes of *Drosophila melanogaster*. Genetics. 20: 192–206.

Beadle, G. W., and E. L. Tatum. 1941. Genetic control of biochemical reactions in *Neurospora*. Proc. Nat. Acad. Sci., Wash. 27: 499–506.

Beale, G. H. 1954. The genetics of *Paramecium aurelia*. Cambridge University Press.

Belling, J. 1931. Chromomeres in liliaceous plants. Univ. Calif. Publ. Bot. 16: 153–70.

Benzer, S. 1955. Fine structure of a genetic region in bacteriophage. Proc. Nat. Acad. Sci., Wash. 41: 344–54.

—— 1957. The elementary units of heredity, in The Chemical Basis of Heredity, pp. 70–93. Baltimore, Johns Hopkins Press.

Boss, J. M. N. 1955. The pairing of somatic chromosomes: a survey. Texas Rep. Biol. Med. 13: 213–21.

Brenner, S. 1957. On the impossibility of all everlapping triplet codes in information transfer from nucleic acid to proteins. Proc. Nat. Acad. Sci., Wash. 43: 687–94.

Bridges, C. B., and K. S. Brehme. 1944. The Mutants of *Drosophila melano-gaster*. Publ. Carnegie Inst. 552.

Buxton, E. W. 1956. Heterokaryosis and parasexual recombination in pathogenic strains of *Fusarium oxysporum*. J. Gen. Microbiol. 15: 133–39.

Calef, E. 1957. Effect on linkage maps of selection of crossovers between closely linked markers. Heredity. 11: 265–79.

Callan, H. G. 1955. Recent work on the structure of cell nuclei, in Fine Structure of Cells, pp. 89–109. Groningen, P. Noordhoff.

Carlson, E. A. 1957. A further analysis of allelism in the dumpy series of *D. melanogaster*. Rec. Genet. Soc. Amer. 26: 363 (Abstr.).

Carter, T. C. 1955. The estimation of total genetical map lengths from linkage test data. J. Genet. 53: 21–28.

Case, M. E., and N. H. Giles. 1957. Evidence from tetrad analysis for both normal crossing over and gene conversion at the *pan-2* locus in *Neurospora crassa*. Rec. Genet. Soc. Amer. 26: 364 (Abstr.).

Chase, M., and A. H. Doermann. 1957. High negative interference over short segments of the genetic structure of bacteriophage T4. Genetics (in press).

Chovnick, A. 1957. Pseudoallelism at the garnet locus in *Drosophila melano-gaster*. Rec. Genet. Soc. Amer. 26: 365 (Abstr.).

Crick, F. H. C. 1955. Discussion of paper by D. Schwartz. J. Cell. Comp. Physiol. 45, Suppl. 2: 184.

—— 1958. On protein synthesis. Symp. Soc. Exp. Biol. 12: 138–63.

Crick, F. H. C., J. S. Griffith, and L. E. Orgel. 1957. Codes without commas. Proc. Nat. Acad. Sci., Wash. 43: 416–21.

Darlington, C. D. 1934. The origin and behaviour of chiasmata. VII. *Zea mays*. Z. indukt. Abst. Vererb. 67: 96–114.

—— 1937. Recent advances in cytology. (2d ed.). London, Churchill.

—— 1939. The evolution of genetic systems. Cambridge University Press.

—— 1956. Natural populations and the breakdown of classical genetics. Proc. Roy. Soc. B. 145: 350–64.

Darlington, C. D., and K. Mather. 1949. The elements of genetics. London, Allen and Unwin.

Davis, B. 1955. Nutritional and enzymatic studies on microbial mutants, pp. 40–58 in Perspectives and Horizons in Microbiology, pp. 40–58. New Brunswick, Rutgers University Press.

Delbrück, M., and G. S. Stent. 1957. On the mechanism of DNA replication, in The Chemical Basis of Heredity, pp. 699–736. Baltimore, Johns Hopkins Press.

Demerec, M. 1928. Mutable characters of *Drosophila virilis* I. Reddish-alpha body character. Genetics. 13: 359–88.

—— 1956. A comparative study of certain gene loci in *Salmonella*. Cold Spring. Harbor. Sym. Quant. Biol. 21: 113–21.

Demerec, M., Z. Hartman, P. E. Hartman, T. Yura, J. S. Gots, H. Ozeki, and S. W. Glover. 1956. Genetic studies with bacteria. Publ. Carnegie Inst. 612.

Dunn, L. C. 1954. The study of complex loci. Caryologia. (Suppl.) 6: 155–66.

—— 1956. Analysis of a complex gene in the house mouse. Cold Spring Harbor Sym. Quant. Biol. 21: 187–94.

Dunn, L. C., and E. Caspari. 1945. A case of neighbouring loci with similar effects. Genetics. 30: 543–68.

Emerson, S. 1956. Notes on the identification of different causes of aberrant tetrad ratios in *Saccharomyces*. C. R. Lab. Carlsberg. 26: 71–86.

Evans, J. V., J. W. B. King, B. L. Cohen, H. Harris, and F. L. Warren. 1956, Genetics of haemoglobin and blood potassium difference in sheep. Nature. London. 178: 849–50.

Eversole, R. A., and E. L. Tatum. 1956. Chemical alteration of crossing over frequency in *Chlamydomonas*. Proc. Nat. Acad. Sci., Wash. 43: 68–73.

Fincham, J. R. S. 1954. Effects of a gene mutation in *Neurospora crassa* relating to glutamic dehydrogenase formation. J. Gen. Microbiol. 11: 236–46.

—— 1957. A modified glutamic acid dehydrogenase as a result of gene mutation in *Neurospora crassa*. Biochem. J. 65: 721–28.

Fincham, J. R. S., and J. B. Boylen. 1957. *Neurospora crassa* mutants lacking arginosuccinase. J. Gen. Microbiol. 16: 438–48.

Fincham, J. R. S., and J. A. Pateman. 1957. Formation of an enzyme through complementary action of mutant "alleles" in separate nuclei in a heterocaryon. Nature. London. 179: 741–42.

Forbes, E. C. 1956. Recombination in the *pro* region in *Aspergillus nidulans*. Microbial Genetics Bull. 13: 9–10.

Gall, J. G. 1956. On the submicroscopic structure of chromosomes. Brookhaven Sym. in Biol. 8: 17–32.

Gans, M. 1953. Étude génétique et physiologique du mutant *z* de *Drosophila melanogaster*. Bull. Biol. Suppl. 38: 1–90.

Giles, N. H. 1951. Studies on the mechanism of reversion in biochemical mutants of *Neurospora*. Cold Spring Harbor Sym. quant. Biol. 16: 283–313.

—— 1956. Forward and backmutation at specific loci in *Neurospora*. Brookhaven Sym. in Biol. 8: 103–23.

Glucksohn-Waelsch, S., H. Ranney, and B. Sisken. 1957. The hereditary transmission of haemoglobin differences in mice. J. Clin. Invest. 36: 753–56.

Goldschmidt, R. B. 1955. Theoretical Genetics. Berkeley, University of California Press.

Goudie, R. B. 1957. Somatic segregation of "inagglutinable" erythrocytes. Lancet, 1957, I: 1333.

Green, M. M. 1954. Pseudoallelism at the vermilion locus in *Drosophila melanogaster*. Proc. Nat. Acad. Sci., Wash. 40: 92–97.

—— 1955a. Phenotypic variation and pseudoallelism at the forked locus in *Drosophila melanogaster*. Proc. Nat. Acad. Sci., Wash. 41: 375–79.

138   WORKS CITED

—— 1955b. Pseudoallelism and the gene concept. Amer. Nat. 89: 65–71.

Green, M. M., and K. C. Green. 1949. Crossing over between alleles at the lozenge locus in *Drosophila melanogaster*. Proc. Nat. Acad. Sci., Wash. 35: 586–91.

Green, M. M., and K. C. Green. 1956. A cytogenetic analysis of the lozenge pseudoalleles in *Drosophila*. Z. indukt. Abst. Vererb. 87: 708–21.

Griffith, F. 1928. The significance of pneumococcal types. J. Hyg., Camb. 27: 113–56.

Grüneberg, H. 1937. Gene doublets as evidence for adjacent small duplications in *Drosophila*. Nature. London. 140: 932.

—— 1952. The genetics of the mouse. The Hague, Nijhoff. Hadorn, E. 1955.

Hadorn, E. 1956. Letalfaktoren in ihre Bedentung für Erbpathologie und Genphysiologie der Entwicklung. Georg Thieme, Stuttgart.

Haldane, J. B. S. 1931. The cytological basis of genetical interference. Cytologia. 3: 54–56.

—— 1932. The time of action of genes and its bearing on some evolutionary problems. Amer. Nat. 66: 5–24.

—— 1941. New pathways in genetics. London, Allen and Unwin.

—— 1954. The biochemistry of genetics. London, Allen and Unwin.

—— 1955. Some alternatives to sex. New Biology. 19: 7–26.

Hartman, P. E. 1956. Linked loci in the control of histidine synthesis in *Salmonella typhimurium*. Publ. Carnegie Inst. 612: 35–62.

—— 1957. Transduction: a comparative review in The Chemical Basis of Heredity, pp. 408–62. Baltimore, Johns Hopkins Press.

Hartman, Z. 1956. Induced mutability in *Salmonella typhimurium*. Publ. Carnegie Inst. 612: 107–20.

Hexter, W. M. 1955. Functional and spatial pseudoallelism at the singed locus in *Drosophila*. Proc. Nat. Acad. Sci., Wash. 41: 921–25.

Horowitz, N. H. 1956. Progress in developing chemical concepts of genetic phenomena. Fed. Proc. 15: 818–22.

Horowitz, N. H. and M. Fling. 1956. Studies on tyrosinase production by a heterocaryon of *Neurospora*. Proc. Nat. Acad. Sci., Wash. 42: 498–501.

Hotchkiss, R. D. 1955. Bacterial transformation. J. Cell. Comp. Physiol. 45 (Suppl.): 1–22.

—— 1957. Discussion of Benzer's paper, in The Chemical Basis of Heredity, pp. 131–32, Baltimore, Johns Hopkins Press.

Howard, A., and S. R. Pelc. 1951. Nuclear incorporation of P^{32} as demonstrated by autoradiographs. Exp. Cell. Res. 2: 178–87.

Hutchinson, J. M. 1958. A first five-marker linkage group identified by mitotic analysis in the asexual *Aspergillus niger*. Microbial Genetics Bull. 15: (in press).

Ingram, V. M. 1957. Gene mutations in human haemoglobin: the chemical difference between normal and sickle-cell haemoglobin. Nature. London. 180: 326–28.

Ishitani, C., Y. Ikeda, and K. Sagaguchi. 1957. Hereditary variation and genetic recombination in Koji-molds (*Aspergillus oryzae* and *Asp. sojae*). VI. Genetic recombination in heterozygous diploids. J. Gen. Appl. Microbiol. (Japan). 2: 401–30.

Ives, P. T., and D. T. Noyes. 1951. A study of pseudoallelism in two multiple allelic series in *Drosophila melanogaster*. Anat. Rec. 111: 565.

Jacob, F., and E. L. Wollman. 1957. Genetic aspects of lysogeny, in The Chemical Basis of Heredity, pp. 468–99. Baltimore, Johns Hopkins Press.

—— 1958. Genetic and physical determinations of chromosomal segments in *Escherichia coli*. Sym. Soc. Exp. Biol. (in press).

James, A. P., and B. Lee-Whiting. 1955. Radiation-induced genetic segregations in vegetative cells in diploid yeast. Genetics. 40: 826–31.

Jensen, K. A., I. Kirk, G. Kølmark, and W. Westergaard. 1951. Chemically induced mutations in Neurospora. Cold Spring Harbor Sym. Quant. Biol. 16: 245–59.

Jinks, J. L. 1952. Heterokaryosis: a system of adaptation in wild fungi. Proc. Roy. Soc. B. 140: 83–106.

Judd, B. H. 1957. Complex pseudoallelism at the white locus in *Drosophila melanogaster*. Rec. Genet. Soc. Amer. 26: 379–89.

Kacser, H. 1956. Molecular organization of genetic material. Science. 124: 151–54.

Käfer, E. 1958. An eight-chromosome map of *Aspergillus nidulans*. Advances in Genetics, 9: 105–45.

Kalckar, H. M., E. P. Anderson, and K. J. Isselbacher. 1956. Galactosemia, a congenital defect in a nucleotide transferase. Proc. Nat. Acad. Sci., Wash. 42: 49–51.

Kaufmann, B. P., and M. R. McDonald. 1956. Organization of the chromosome. Cold Spring Harbor Sym. Quant. Biol. 21: 233–46.

Komai, T. 1950. Semi-allelic genes. Amer. Nat. 84: 381–92.

Koske, T., and J. Maynard-Smith. 1954. Genetics and cytology of *Drosophila subobscura* J. Genet. 52: 521–41.

Kurnick, N. B., and I. H. Herskowitz. 1952. The estimation of polyteny in *Drosophila* salivary gland nuclei based on determination of DNA content. J. Cell. Comp. Physiol. 39: 281–99.

Laughnan, J. R. 1955. Structural and functional aspects of the A^6 complex in maize. Proc. Nat. Acad. Sci., Wash. 41: 78–84.

Lederberg, E. 1952. Allelic relationships and reverse mutation in *Escherichia coli*. Genetics. 37: 469–83.

Lederberg, J. 1947. Gene recombination and linked segregations in *Escherichia coli*. Genetics. 32: 505–25.

—— 1951. Genetic studies with Bacteria, in Genetics in the 20th Century, pp. 263–89. New York, Macmillan.

—— 1955. Recombination mechanisms in bacteria. J. Cell. Comp. Physiol. 45 (Suppl. 2): 75–107.

Lederberg, J., and E. Lederberg. 1956. Infection and heredity, in Cellular mechanisms of differentiation and growth, pp. 101–24. Princeton University Press.

Leupold, U. 1957. Physiologisch-genetische Studien an adenin-abhängingen Mutanten von *Schizosaccharomyces pombe*. Ein Beitrag zum Problem der Pseudoallelie. Schweiz. Z. allg. Path. Bakt. 20: (in press).

Levine, R. P. 1955. Chromosome structure and the mechanism of crossing over. Proc. Nat. Acad. Sci., Wash. 41: 727–30.

Levinthal, C. 1954. Recombination in phage T2; its relationship to heterozygosis and growth. Genetics. 39: 169–84.

—— 1956. The mechanism of DNA replication and genetic recombination in Phage. Proc. Nat. Acad. Sci., Wash. 42: 394–404

Lewis, D. 1954. Comparative incompatibility in Angiosperms and Fungi. Adv. Genet. 6: 235–85.

Lewis, E. B. 1945. The relation of repeats to position effect in *Drosophila melanogaster*. Genetics. 30: 137–66.

—— 1951. Pseudoallelism and gene evolution. Cold Spring Harbor Sym. Quant. Biol. 16: 151–72.

—— 1952. The pseudoallelism of white and apricot in *Drosophila melanogaster*. Proc. Nat. Acad. Sci., Wash. 38: 953–56.

—— 1954. Pseudoallelism and the gene concept. Caryologia. 6 (Suppl.): 100-105.

—— 1955. Some aspects of position pseudoallelism. Amer. Nat., 89: 75–89.

Lindegren, C. C. 1953. Gene conversion in *Saccharomyces*. J. Genet. 51: 625-37.

Lindegren, C. C., and G. Lindegren. 1937. Non-random crossing over in *Neurospora*. J. Hered. 28: 105-13.

Lwoff, A. 1953. Lysogeny. Bact. Rev. 17: 269–337.

McClintock, B. 1956. Controlling elements and the gene. Cold Spring Harbor Sym. Quant. Biol. 21: 197–216.

Mackendrick, M. E. 1953. Further examples of crossing over between alleles of the *w* series. Drosophila Information Service. 27: 10.

Mackendrick, M. E., and G. Pontecorvo. 1952. Crossing over between alleles at the *w* locus in *Drosophila melanogaster*. Experientia. 8: 390.

Martin-Smith, C. A. 1958. The *ad9* series in *Aspergillus nidulans*. Microbial Genetics Bull. 15: 20–21.

Mather, K. 1943. Polygenic inheritance and natural selections. Biol. Rev. 18: 32-64.

—— 1948. Nucleus and cytoplasm in differentiation. Sym. Soc. Exp. Biol. 2: 196–216.

Mazia, D. 1954. The particulate organisation of the chromosome. Proc. Nat. Acad. Sci., Wash. 40: 521–27.

Mitchell, H. K. 1957. Crossing over and gene conversion in *Neurospora*, pp. 94-113 in The Chemical Basis of Heredity, pp. 94–113. Baltimore, Johns Hopkins Press.

Mitchell, M. B. 1955. Aberrant recombination of pyridoxine mutants in *Neurospora*. Proc. Nat. Acad. Sci., Wash. 41: 215–20.

—— 1956. A consideration of aberrant recombination in Neurospora. C. R. Lab. Carlsberg, Ser. Physiol. 26: 285–98.

Morse, M. L., E. M. Lederberg, and J. Lederberg. 1956. Transductional heterogenotes in *Escherichia coli*. Genetics, 41: 758–79.

Muller, H. J. 1916. The mechanism of crossing over. Amer. Nat. 50: 193–221; 284–305; 350–66; 421–34.

—— 1926. The gene as the basis of life. Proc. 4th Int. Congr. Plant. Sci., Ithaca. 1: 892–921 (published 1929).

—— 1935. On the dimensions of chromosomes and genes in Dipteran salivary chromosomes. Amer. Nat. 69: 405–11.

—— 1938. The position effect as evidence of the localisation of the immediate products of gene activity. Proc. 15th Int. Physiol. Congr. (Moscow). 587–89.

—— 1947a. The gene. Proc. Roy. Soc. B. 134: 1–37.

—— 1947b. Genetic fundamentals: the dance of the genes, in Genetics, Medicine and Man, pp. 35–65. Cornell University Press.

Myers, J. W., and E. A. Adelberg. 1954. The biosynthesis of isoleucine and valine. I. Enzymatic transformation of the hydroxy acid precursor to the keto acid precursor. Proc. Nat. Acad. Sci., Wash. 40: 493–99.

Neuhaus, M. E. 1939. A cytogenetic study of the Y-chromosome of *Drosophila melanogaster*. J. Genet. 37: 229–54.

Newmayer, D. 1957. Arginine synthesis in *Neurospora crassa*: Genetic studies. J. Gen. Microbiol. 16: 449–62.

Ogur, M., and Rosen, G. U., 1950. The nuclei acids of plant tissues. Arch. Biochem. 25: 262–76.

Ozeki, H. 1956. Abortive transduction in purine-requiring mutants of *Salmonella typhimurium*, in Genetic studies with bacteria, pp. 97–106. Publ. Carnegie Inst. 612.

Pauling, L., H. A. Itano, S. J. Singer, and I. C. Wells. 1949. Sickle-cell anaemia, a molecular disease. Science. 110: 543–48.

Penrose, L. S., and R. Penrose. 1957. A self-reproducing analogue. Nature. London. 179: 1183.

Perkins, D. D. 1953. The detection of linkage in tetrad analysis. Genetics. 38: 187–97.

Perkins, D. D. 1955. Tetrads and crossing over. J. Cell. Comp. Physiol. 45 (Suppl. 2): 119–49.

—— 1956. Crossing over in a multiply marked chromosome arm of *Neurospora*. Microbial Genetics Bull. 13: 22–23.

Plaut, W., and D. Mazia. 1956. The distribution of newly synthesised DNA in mitotic division. J. Biophys. Biochem. Cyt. 2: 573–88.

Pollister, A. W., H. Swift, and M. Alfert. 1951. Studies on the desoxypentose nucleic acid content of animal nuclei. J. Cell. Comp. Physiol. 38 (Suppl. 1): 101–19.

Pontecorvo, G. 1944. Structure of heterochromatin. Nature. London. 153: 365
—— 1946. Genetic systems based on heterocaryosis. Cold Spring Harbor Sym Quant. Biol. 11: 193–201.
—— 1950. New fields in the biochemical genetics of microorganisms. Biochem Soc. Symp. 4: 40–50.
—— 1952. The genetic formulation of gene structure and action. Advances i Enzymology. 13: 121–49.
—— 1953. Non-random distribution of multiple mitotic crossing over. Natur London. 170: 204.
—— 1954. Mitotic recombination in the genetic system of filamentous fung Caryologia. Suppl. 6: 192–200.
—— 1955. Gene structure and action in relation to heterosis. Proc. Roy. Soc. B 144: 171–77.
—— 1956a. Allelism. Cold Spring Harbor Sym. Quant. Biol. 21: 171–74.
—— 1956b. The parasexual cycle in fungi. Ann. Rev. Microbiol. 10: 393–400.
Pontecorvo, G., and E. Käfer. 1954. Maps of a chromosome region in Asper gillus nidulans based on mitotic and meiotic crossing over. Heredity. 8: 43. (Abst.).
Pontecorvo, G., and E. Käfer. 1956. Mapping the chromosomes by means o mitotic recombination. Proc. Roy. Phys. Soc. Edinburgh 25: 16–20.
Pontecorvo, G., and E. Käfer. 1958. Genetic analysis by means of mitotic re combination. Advances in Genetics, 9: 71–104.
Pontecorvo, G., and J. A. Roper. 1952. Genetic analysis without sexual repr duction by means of polyploidy in Aspergillus nidulans. J. Gen. Microbio 6: vii (Abst.).
Pontecorvo, G. and J. A. Roper. 1956. The resolving power of genetic analysis Nature. London. 178: 83–84.
Pontecorvo, G., J. A. Roper, and E. Forbes. 1953. Genetic recombination with out sexual reproduction in Aspergillus niger. J. Gen. Microbiol. 8: 198–210
Pontecorvo, G., J. A. Roper, L. M. Hemmons, K. D. Macdonald, and A. W. J Bufton. 1953. The genetics of Aspergillus nidulans. Advances in Genetics 5: 141–238.
Pontecorvo, G., and G. Sermonti. 1954. Parasexual recombination in Peni cillium chrysogenum. J. Gen. Microbiol. 11: 94–104.
Pontecorvo, G., E. Tarr-Gloor, and E. Forbes. 1954. Analysis of mitotic recom bination in Aspergillus nidulans. J. Genet. 52: 226–37.
Pritchard, R. H. 1955. The linear arrangement of a series of alleles in Asper gillus nidulans. Heredity. 9: 343–71.
—— 1958. Localised negative interference in Aspergillus nidulans. Microbia Genetics Bull. 15: 22–24.
Raffel, D., and H. J. Muller. 1940. Position effect and gene divisibility consid ered in connection with three strikingly similar scute mutations. Genetics 25: 541–83.
Rhoades, M. M. 1950. Meiosis in Maize. J. Heredity. 41: 59–67.

is, H. 1957. Chromosome structure, in The Chemical Basis of Heredity, pp. 23–62. Baltimore, Johns Hopkins Press.

oman, H. 1956. Studies of genes mutation in *Saccharomyces*. Cold Spring Harbor Sym. Quant. Biol. 21: 175–83.

oper, J. A. 1950. A search for linkage between genes determining vitamin requirements. Nature. London. 166: 956.

—— 1952. Production of heterozygous diploids in filamentous fungi. Experientia. 8: 14–15.

oper, J. A., and R. H. Pritchard. 1955. The recovery of the reciprocal products of mitotic crossing over. Nature. London. 175: 639.

yan, F., and J. Lederberg. 1946. Reverse mutation and adaptation in leucineless *Neurospora*. Proc. Nat. Acad. Sci., Wash. 32: 165–73.

t. Lawrence, P. 1956. The *q* locus of *Neurospora crassa*. Proc. Nat. Acad. Sci., Wash. 42: 189–94.

ansome, E. 1949. Spontaneous mutation in standard and "Gigas" forms of *Penicillium notatum*. Trans. Brit. Mycol. Soc. 32: 305–14.

chmitt, F. O. 1956. Chromosomes, genes, and macromolecular systems. Nature. London. 177: 503–5.

chröedinger, E. 1943. What is life? Cambridge University Press.

chwartz, D. 1955. Studies on crossing over in maize and *Drosophila*. J. Cell. Comp. Physiol. 45 (Suppl. 2.): 171–88.

ermonti, G. 1955. Genetics of *Penicillium chrysogenum*. II. Segregation and recombination from a heterozygous diploid. Rend. Ist Sup. Sanità (English Edition). 17: 231–43.

ilver, W. S., and W. D. McElroy. 1954. Enzyme studies on Nitrate and Nitrite mutants of *Neurospora*. Arch. Biochem. 51: 379–94.

ingleton, J. R. 1953. Chromosome morphology and the chromosome cycle in the ascus of *Neurospora crassa*. Amer. J. Bot. 40: 124–44.

latis, H. M., and D. A. Willermet. 1953. The miniature complex in *Drosophila melanogaster*. Genetics. 39: 45–58.

lizynski, B. M. 1945. "Ectopic" pairing and the distribution of heterochromatin in the X-chromosome of salivary gland nuclei of *Drosophila melanogaster*. Proc. Roy. Soc. Edinburg B. 114–19.

—— 1955. Chiasmata in the male mouse. J. Genet. 53: 597–605.

tadler, L. J. 1951. Spontaneous mutation in maize. Cold Spring Harbor Sym. Quant. Biol. 16: 49–63.

teffensen, D. 1955. Breakage of chromosomes in *Tradescantia* with a calcium deficiency. Proc. Nat. Acad. Sci., Wash. 41: 155–60.

tent, G. S. 1958. Mating in the reproduction of bacterial viruses. Adv. Virus Res. 5: (in press).

tern, C. 1936. Somatic crossing over and segregation in *Drosophila melanogaster*. Genetics, 21: 625–730.

tern, C., and E. W. Schaeffer. 1943. On wild-type isoalleles in *Drosophila melanogaster*. Proc. Nat. Acad. Sci., Wash. 29: 361–67.

Streisinger, G., and N. C. Franklin. 1956. Mutation and recombination at the host range genetic region of phage T2. Cold Spring Harbor Sym. Quant. Biol. 21: 103–9.

Strickland, W. N. 1958*a*. Abnormal tetrads in *Aspergillus nidulans*. Proc. Roy. Soc. B. 533–42.

—— 1958*b*. An analysis of interference in *Aspergillus nidulans*. Proc. Roy. Soc. B. 82–101.

Sturtevant, A. H. 1951*a*. The relation of genes and chromosomes, in Genetics in the 20th Century, pp. 101–10. New York, Macmillan.

—— 1951*b*. A map of the fourth chromosome in *Drosophila melanogaster* based on crossing over in triploid females. Proc. Nat. Acad. Sci., Wash. 37: 405–7.

—— 1955. Evaluation of recombination theory. J. Cell. Comp. Physiol. 45 (Suppl. 2): 237–42.

Tanaka, Y. 1953. Genetics of the silkworm. Advances in Genetics. 5: 239–316.

Taylor, J. H. 1953. Autoradiographic detection of incorporation of P^{32} into chromosomes during meiosis and mitosis. Exp. Cell Res. 4: 169–79.

—— 1957. The time and mode of duplication of chromosomes. Amer. Nat. 91: 209–21.

Tsujita, M., and B. Sakaguchi. 1953. Studies on the semi-allelic *E*-series in the silkworm. Rep. Nat. Inst. Genet. (Japan). 3: 20–26.

Vendreli, R. 1955. The deoxyribonucleic acid content of the nucleus, in The Nucleic Acids, pp. 155-80. New York, Academic Press.

Waddington, C. H. 1956. Principles of Embryology. London, Allen and Unwin.

Watson, J. D., and F. H. C. Crick. 1953. Genetical implications of the structure of deoxyribose nucleic acid. Nature. London. 171: 964.

Watson, J. D., and G. Maalöe. 1953. Nucleic acid transfer from parental to progeny bacteriophage. Biochem. Biophys. Acta. 10: 432–42.

Westergaard, M. 1957. Chemical mutagenesis in relation to the concept of the gene. Experientia. 13: 224–34.

Whitehouse, H. L. K. 1956. The use of loosely linked genes to estimate chromatid interference by tetrad analysis. C. R. Lab. Carlsberg. 26: 407–22.

Whiting, P. W. 1950. Blood group symbols and genes: some thoughts on allelism, pleiotropy and dominance. J. Hered. 41: 55–56.

Whittinghill, M. 1955. Crossover variability and induced crossing over. J. Cell. Comp. Physiol. 45(Suppl. 2): 189–220.

Winge, Ø. 1955. On interallelic crossing over. Heredity. 9: 373–84.

Winge, Ø., and C. Roberts. 1954. On tetrad analyses apparently inconsistent with Mendelian law. Heredity, 8: 295–305.

Wollman, E., F. Jacob, and W. Hayes. 1956. Conjugation and genetic recombination in *Escherichia coli* K. 12. Cold Spring Harbor Sym. Quant. Biol. 21: 141–62.

Wright, Sewall. 1953. Gene and organism. Amer. Nat. 87: 5–18.

Yanofski, C., and D. M. Bonner. 1955. Gene interaction in tryptophane synthetase formation. Genetics. 40: 761–69.

Yu, C. P., and T. S. Chang. 1948. Further studies on the inheritance of anthocyanin pigmentation in Asiatic cotton. J. Genet. 49: 46–56.

Zinder, N. D., and J. Lederberg. 1952. Genetic exchange in *Salmonella*. J. Bact. 64: 679–99.